AF689816

私人轎車與中國

Personal Cars and China

CHINESE ACADEMY OF ENGINEERING

NATIONAL RESEARCH COUNCIL
OF THE NATIONAL ACADEMIES

THE NATIONAL ACADEMIES PRESS
Washington, D.C.
www.nap.edu

THE NATIONAL ACADEMIES PRESS • 500 Fifth Street, N.W. • Washington, D.C. 20001

NOTICE: The project that is the subject of this report was approved by the Governing Board of the National Research Council, whose members are drawn from the councils of the National Academy of Sciences, the National Academy of Engineering, and the Institute of Medicine. The members of the committee responsible for the report were chosen for their special competences and with regard for appropriate balance.

This study was supported by grants from the Ford Motor Company, DaimlerChrysler, U.S. Department of Energy, U.S. National Research Council, U.S. National Academy of Engineering, Shanghai Automotive Industry Corporation, China FAW Group Corporation, Dongfeng Motor Corporation, and Chinese Academy of Engineering. Any opinions, findings, conclusions, or recommendations expressed in this publication are those of the author(s) and do not necessarily reflect the views of the organizations or agencies that provided support for the project.

International Standard Book Number 0-309-08492-X

Additional copies of this report are available from the National Academies Press, 500 Fifth Street, N.W., Lockbox 285, Washington, D.C. 20055; (800) 624-6242 or (1-202) 334-3313 (in the Washington metropolitan area); Internet, http://www.nap.edu

Copyright 2003 by the National Academy of Sciences. All rights reserved.

Printed in the United States of America.

THE CHINESE ACADEMY OF ENGINEERING

The **Chinese Academy of Engineering (CAE)** is China's foremost academic and advisory institution in engineering and technological sciences. It was founded in June 1994 and now has 611 members. As a national institution, CAE has the following missions: to promote the progress of engineering and technological sciences, foster the growth of outstanding talents in close collaboration with the engineering and technological community, and enhance international cooperation in order to facilitate sustainable economic and social development in China. Prof. Xu Kuangdi is president of the Chinese Academy of Engineering. Dr. Song Jian is the former president (June 1998–June 2002).

THE NATIONAL ACADEMIES

The **National Academy of Sciences** is a private, nonprofit, self-perpetuating society of distinguished scholars engaged in scientific and engineering research, dedicated to the furtherance of science and technology and to their use for the general welfare. Upon the authority of the charter granted to it by the Congress in 1863, the Academy has a mandate that requires it to advise the federal government on scientific and technical matters. Dr. Bruce M. Alberts is president of the National Academy of Sciences.

The **National Academy of Engineering** was established in 1964, under the charter of the National Academy of Sciences, as a parallel organization of outstanding engineers. It is autonomous in its administration and in the selection of its members, sharing with the National Academy of Sciences the responsibility for advising the federal government. The National Academy of Engineering also sponsors engineering programs aimed at meeting national needs, encourages education and research, and recognizes the superior achievements of engineers. Dr. Wm. A. Wulf is president of the National Academy of Engineering.

The **Institute of Medicine** was established in 1970 by the National Academy of Sciences to secure the services of eminent members of appropriate professions in the examination of policy matters pertaining to the health of the public. The Institute acts under the responsibility given to the National Academy of Sciences by its congressional charter to be an adviser to the federal government and, upon its own initiative, to identify issues of medical care, research, and education. Dr. Harvey V. Feinberg is president of the Institute of Medicine.

The **National Research Council** was organized by the National Academy of Sciences in 1916 to associate the broad community of science and technology with the Academy's purposes of furthering knowledge and advising the federal government. Functioning in accordance with general policies determined by the Academy, the Council has become the principal operating agency of both the National Academy of Sciences and the National Academy of Engineering in providing services to the government, the public, and the scientific and engineering communities. The Council is administered jointly by both Academies and the Institute of Medicine. Dr. Bruce M. Alberts and Dr. Wm. A. Wulf are chairman and vice chairman, respectively, of the National Research Council.

THE NATIONAL ACADEMIES

Advisers to the Nation on Science, Engineering, and Medicine

www.national-academies.org

COMMITTEE ON THE FUTURE OF PERSONAL TRANSPORT VEHICLES IN CHINA

Dale COMPTON, Cochairman (NAE)
Lillian M. Gilbreth Distinguished Professor of Industrial Engineering
School of Industrial Engineering
Purdue University

GUO Konghui, Cochairman (CAE)
Director, State Key Lab for Automobile Dynamic Simulation
Professor, Jilin University

Feng AN
Consultant
Center for Transportation Research
Argonne National Laboratory

CHENG Zhenbiao
Chief Engineer
Science and Technology Information Research Institute
Dongfeng Motor Corporation

Elisabeth DRAKE
Associate Director for New Energy Technology (retired)
Energy Laboratory
Massachusetts Institute of Technology

Patrick F. FLYNN
Vice President, Research (retired)
Cummins Engine Company

Ralph GAKENHEIMER
Professor of Urban Planning (retired)
Department of Urban Studies and Planning
Massachusetts Institute of Technology

Gregory K. INGRAM
Director
Operations Evaluation Department
World Bank

LI Gang
Former President, Chinese National Automotive Industry Corporation
Senior Project Adviser, Chinese Academy of Engineering

LU Ximing
Director
Shanghai City Comprehensive Transportation Planning Institute

QIU Yansheng
Deputy Chief Engineer, Professor
Research Institute of Petroleum Processing (RIPP)

SHAO Min
Deputy Director, Associate Professor
Center for Environmental Sciences
Peking University

Daniel SPERLING
Professor of Civil and Environmental Engineering and Environmental Studies
Director, Institute of Transportation Studies
University of California, Davis

SUN Zhenhua
Senior Engineer
Technique and Quality Department
Shanghai Automotive Industry Corporation (SAIC)

Michael P. WALSH
Consultant
Arlington, Virginia

ZHAO Junming
Senior Engineer
Changchun Automotive Research Institute (CARI)

Ex officio Members

ZHU Gaofeng
Vice President (June 1994–June 2002)
Chinese Academy of Engineering

Harold FORSEN
Foreign Secretary
National Academy of Engineering

Staff

Michael GREENE, Study Director (NRC)
Division of Policy and Global Affairs
National Research Council

LI Renhan, Study Director (CAE)
Director
Division Office of Mechanical and Vehicle Engineering
Chinese Academy of Engineering

ZHENG Xiaoguang
Director
Department of International Cooperation
Division of International Organization
Chinese Academy of Engineering

Vivienne CHIN
Program Office
National Academy of Engineering

Preface

In mid-1999 representatives of the Chinese Academy of Engineering (CAE) visited the U.S. National Academy of Engineering (NAE) and National Research Council (NRC) to explore the prospects for collaboration among the institutions on a study of the future of the personal car in China. This is the second instance of formal collaboration between the CAE and the National Academies. The outcome of the first was a study entitled *Cooperation in the Energy Futures of China and the United States,* published in 2001, in which the Chinese Academy of Sciences also participated.

The National Research Council, the operating arm of the three National Academies—the National Academy of Sciences, National Academy of Engineering, and Institute of Medicine—has been producing independent advisory reports at the request of the U.S. government and other government and donor organizations since 1916 (the parent organization, the National Academy of Sciences, received its congressional charter in 1863). The National Academy of Engineering was created in 1964, and carries out advisory studies and other programs on matters of technology. The Chinese Academy of Engineering has been in existence since 1994, and is developing a role as adviser to its government that parallels that of the NAE and NRC.

Although the nominal topic of this report—the Chinese automotive industry and the future of personal cars in China—is specific to China, many of the issues examined also are relevant to the United States and other countries. For example, the higher polluting emissions that will accompany the proliferation of cars in China are predicted to have global implications for climate change. Moreover, an expanding Chinese automotive fleet will increase the world demand for petroleum, and raises the possibility of higher international prices as China becomes a major petroleum importer. Those and the other problems that will confront China as it continues to develop its transportation system—such as congestion, more accidents, undesirable changes in land use, and urban decentralization—must be addressed by any nation that expects to see its motor vehicle population grow significantly.

This study could not examine the extent of global problems that may ensue from a large increase in motorization worldwide, but the committee felt very strongly that the impacts, both positive and negative, of any such increase should be thoroughly explored. Essentially all of the issues identified in this report as being critical to China will arise on a global scale and will be that much more difficult to manage. We urge that such a study be undertaken at the earliest possible moment.

Some important issues identified that are specific to China could not be explored in detail within the present study. These include the impact of motorization on inequity among various segments of the population and the cost and financing of the new infrastructure that will accompany increased motorization in China. An in-depth study of these issues, however, would require an examination of regional and national economic and social development, and the committee was not prepared to examine the many ramifications of this subject.

Similarly, the report devotes less space to social change and changes in urban form than to automotive and fuel-related technologies. Although the report does observe that increases in motorization will lead to decentralization of both jobs and residences in Chinese cities, with some illustrative calculations for Shanghai, answering important questions about the economic impacts and how the transportation infrastructure is to be financed involves issues of public revenues and expenditures that were well beyond the charge to the committee. Nevertheless, these issues are of immense importance to China and deserve further study.

As the goals and priorities of the Chinese automotive industry and the Chinese government evolve, it is clear that the many ways to achieve them will come to light, but none certain of success and all with palpable risks. Naturally, these conflicts were reflected in differences of opinion among members of the committee; however, the line of advocacy on each side of these issues rarely coincided with the nationalities of the committee members. It was a committee of individual experts who, taken together, represented all points of view. Each side learned from the other, and the collaboration has strengthened both of the collaborating institutions. We hope that this report will be viewed as a useful contribution to policy making by the automotive industries and governments of all countries and will serve as an important addition to the literature of science and technology policy. We were honored to serve as cochairmen of this distinguished committee. And we compliment the members of the committee for their diligence and efforts throughout this study to ensure that it properly reflected the challenges and opportunities of this period of dynamic change for the Chinese automotive industry and the people and government of China.

W. Dale Compton
U.S. National Academy of Engineering
U.S. National Research Council

Guo Konghui
Chinese Academy of Engineering

Acknowledgments

The committee would like to thank the Ford Motor Company, DaimlerChrysler, U.S. Department of Energy, U.S. National Research Council (NRC), U.S. National Academy of Engineering, Shanghai Automotive Industry Corporation, China FAW Group Corporation, Dongfeng Motor Corporation, and Chinese Academy of Engineering (CAE) for financial and other support for the study. Staff of the National Research Council's Transportation Research Board and Board on Energy and Environmental Systems, particularly Stephen Godwin and James Zucchetto, helped to launch the study, and Lester Hoel and Harold Forsen of the National Academy of Engineering traveled to Shanghai to help formulate the memorandum of understanding. Zhu Gaofeng of the Chinese Academy of Engineering made many valuable contributions during the deliberations. The study directors—Michael Greene, Zheng Xiaoguang, and Li Renhan—provided essential logistical support for the committee and valuable contributions during the deliberations and the drafting of the report. Project consultant Sabra Bissette Ledent ably edited the report.

This report has been reviewed in draft form by individuals chosen for their diverse perspectives and technical expertise, in accordance with procedures approved by the NRC's Report Review Committee and the CAE's Committee for Consultative Projects. The purpose of these independent reviews is to provide candid and critical comments that will assist the institutions in making their published report as sound as possible and to ensure that the report meets institutional standards for objectivity, evidence, and responsiveness to the study charge. The review comments and draft manuscript remain confidential to protect the integrity of the deliberative process.

We wish to thank the following for their review of this report for the NRC: William Agnew, General Motors (retired); Scott Elliott, Los Alamos National Laboratory; D. Gale Johnson, University of Chicago; He Kebin, Tsinghua University; Zhi Liu, World Bank; Michael Meyer, Georgia Institute of Technology; Roberta Nichols, Ford Motor Company (retired); Joseph Norbeck, University of California, Riverside; Jerrold Voss, Ohio State University; and Martin Wachs, University of California, Berkeley. We also wish to thank the reviewers for the Chinese Academy of Engineering: Hu Xinmin, China National Automotive Industrial Corporation; Li Jingsheng, Department of International Cooperation, China Machinery Industry Federation; Lu Zhiqiang, Development Research Center of the State Council of the Peoples Republic of China; Quan Yongshen, Beijing Transportation Research Institute; Tang Xiaoyan, Center for Environmental Sciences of Peking University; Xu Shoubo, Scientific Committee of State Planning Commission of the Peoples Republic of China; and Zhang Xingye, Automotive Society of China.

Although the reviewers just listed provided many constructive comments and suggestions, they were not asked to endorse the conclusions or recommendations of the report, nor did they see the final draft of the report before its release. The NRC review of this report was overseen by Lester Hoel, University of Virginia, and Morris Tannenbaum, AT&T Corporation (retired), who were appointed by the National Research Council to make certain that an independent examination of this report was carried out in accordance with NRC and CAE procedures and that all review comments were carefully considered. For the Chinese Academy of Engineering, the Committee for Consultative Projects carried out this responsibility. Responsibility for the final content of this report rests entirely with the authoring committee, the CAE, and the NRC.

Finally, we owe thanks to those who have translated the English version of the report into Chinese.

Contents

LIST OF FIGURES, TABLES, AND BOXES

Figures

Tables

Boxes

ACRONYMS AND ABBREVATIONS

A5	Automotive five gear
AB	advanced body
ABS	antilock braking system
AC	alternating current
ACEA	European Automobile Manufacturers Association
A/F	air and fuel
ASM	acceleration simulation mode (test)
atm	atmosphere
bbl	barrel, barrels
bhp	brake horsepower
BMEP	brake mean effective pressure
BSFC	brake-specific fuel consumption
Btu	British thermal unit
CAE	Chinese Academy of Engineering
CAFE	Corporate Average Fuel Economy
CARB	California Air Resources Board
CBM	coal-bed methane
CDH	Colorado Department of Health
CFC	chlorofluorocarbon
CFPP	cold filter plugging point
CH_4	methane
CI	compression ignition
CIDI	compression ignition direct injection
CNG	compressed natural gas
CNPC	China National Petroleum Corporation
CO	carbon monoxide
CO_2	carbon dioxide
CV	conventional vehicle
CVCC	compound vortex-controlled combustion
CVS	constant volume sampler
CVT	continuously variable transmission
DC	direct current
DMC	Dongfeng Motor Corporation
DME	dimethyl ether
DPI	direct public investment
ECD	Energy Conversion Devices Inc.
EGR	exhaust gas recirculation

EMAT	electro-mechanical automatic transmission
EPA	Environmental Protection Agency (U.S.)
EU	European Union
EUCAR	European Council for Automotive R&D
FAW	First Auto Works
FC	fuel cell
FP	fine particles
F-T	Fischer-Tropsch
FTP	Federal Test Procedure (U.S.)
g	gram
GATS	General Agreement on Trade in Services
GATT	General Agreement on Tariffs and Trade
GCP	Gross City Product
GDI	gasoline direct injection
GDP	gross domestic product
GHG	greenhouse gas
GIS	geographic information system
GJ	gigajoule
GM	General Motors
GPS	global positioning system
GWP	global warming potential
H_2	hydrogen
HC	hydrocarbon
HCCI	homogeneous charge compression ignition
HCFC	hydrochlorofluorocarbon
HEV	hybrid electric vehicle
HFC	hydrofluorocarbon
hp	horsepower
HVAC	heating, ventilating, air-conditioning
I	iodine
IARC	International Agency for Research on Cancer
ICE	internal combustion engine
I/M	inspection and maintenance
IPCC	International Panel on Climate Change
IRF	International Road Federation
ITS	intelligent transportation system
J	joule
JAMA	Japanese Automobile Manufacturers Association

KAMA	Korean Automobile Manufacturers Association
kg	kilogram
kl	kiloliter
km	kilometer
KOH	potassium hydroxide
kPa	kilopascal
kph	kilometers per hour
kW	kilowatt
kWh	kilowatt-hour
lb	pound
lb/hp-hr	pounds mass per horsepower-hour
LEV	low-emission vehicle
LHV	lower heating value
LNG	liquefied natural gas
LPG	liquefied petroleum gas
m^3	cubic meter
M5	manual five gear
mbd	million barrels per day
MeOH	methanol
Mg	magnesium
MITI	Ministry of International Trade and Industry (Japan)
MJ	megajoule
MMT	methyl cyclopentadienyl manganese tricarbonyl
MMT	million metric tons
MON	Motor Octane Number
mpg	miles per gallon
mph	miles per hour
MPV	multipurpose vehicle
MTBE	methyl tertiary-butyl ether
MW	megawatts
μg	microgram
μm	micron, micrometer
MWh	megawatt-hour
NASA	National Aeronautics and Space Administration (U.S.)
NDIR	nondispersive infrared
NH_3	ammonia
NHDTG	National Heavy-duty Truck Group (China)
NiMH	nickel metal hydride
NMHC	nonmethane hydrocarbon

NMVOC	nonmethane volatile organic compound
NO_2	nitrogen dioxide
NO_x	nitrogen oxides
N_2O	nitrous oxide
NRC	National Research Council
O_3	ozone
ODS	ozone-depleting substance
OECD	Organisation for Economic Co-operation and Development
OEM	original equipment manufacturer
PAH	polycyclic aromatic hydrocarbon
PEM	proton exchange membrane
PFC	perfluorocarbon
PM	particulate matter
PNGV	Partnership for a New Generation of Vehicles (U.S.)
ppm	parts per million
R&D	research and development
RMB	renminbi
RON	Research Octane Number
rpm	revolutions per minute
RVP	Reid vapor pressure
s	second
SAE	Society of Automotive Engineers
SAIC	Shanghai Automotive Industry Corporation
SCR	selective catalytic reduction
SF_6	sulfur hexafluoride
SI	spark ignition
SINOPEC	China Petroleum and Chemical Corporation
SOFC	solid oxide fuel cell
SPS	sanitary and phytosanitary
SUV	sport-utility vehicle
TAIC	Tianjin Automotive Industry Group Corporation
TBT	Technical Barrier to Trade [Agreement]
TOE	tons of oil equivalent
TRIMS	[Agreement on] Trade-Related Investment Measures
TRIPS	Trade-Related Aspects of Intellectual Property Rights

UHC	unburned hydrocarbons
USABC	U.S. Advanced Battery Consortium
USCAR	U.S. Council for Automotive Research
USEPA	U.S. Environmental Protection Agency
V	volt
VMT	vehicles miles traveled
VOC	volatile organic compound
W	watt
WTO	World Trade Organization

Executive Summary

In 1991 the Chinese government published its eighth five-year plan (1991–1995), which designated the automotive industry as a "pillar industry" that would drive the economy in the twenty-first century. Accordingly, in its most recent five-year plan for the Chinese automotive industry (2001–2005), the government stated that its immediate goal is to produce over 1 million cars a year.

At present, China has relatively few motor vehicles per capita, and, of the cars on its highways, few are privately owned. In 2001 China had 18 million vehicles, of which 5 million were cars. If China's number of motor vehicles per capita were comparable to the world average, its fleet would have to total 160 million, with 10 million new and replacement vehicles acquired each year.

The 2001–2005 plan for the automotive industry calls for its massive restructuring—from 118 individual manufacturers to 3 large automotive groups and from several hundred parts suppliers to 5–10 large supplier groups. The industry also is encouraged to produce, independently of foreign manufacturers, a Chinese economy car, utilizing a 1.3-liter engine and meeting Chinese emissions and fuel economy standards, that could be purchased for less than RMB80,000 ($9,800). Investments in highways, oil and gas pipelines, and other transportation infrastructure are expected to accompany the expansion of car ownership.

Assuming continued economic growth, it is highly likely that China's vehicle fleet will grow rapidly. Although such a rapid growth will bring many benefits to China, it also will present its social, environmental, and economic systems with serious challenges.

In mid-1999 representatives of the Chinese Academy of Engineering (CAE) visited the U.S. National Research Council (NRC) to explore the prospects for collaboration between the two institutions on a study of the future of the personal car in China. The study was to suggest strategies for developing a Chinese national car, as described in China's five-year plan for the automotive industry, and the role of such a car in the national transportation system. It would take into account China's social development, opportunities for cooperation between government and industry, and the impact of a large increase in the number of private cars on sustainable development. The study also was to examine the various options that might be available to mitigate problems such as increased congestion, pollution, and energy consumption. In 2001 the study committee, composed of an equal number of Chinese and American experts, began work.

THE IMPACT OF RAPID MOTORIZATION ON CHINA'S CITIES AND ECONOMY

Whether from domestic or overseas sources, a rapid increase in the number of cars in China will produce both benefits and liabilities. In the short term, a more mobile population will have greater choices in housing, employment, shopping, and leisure. But the experience of other countries suggests that in urban areas, in the absence of government intervention, poorer air quality, more auto accidents, and increased congestion will negatively affect all urban residents. In the longer term, the developed areas of cities may expand as populations and their employers move outward from the city center, away from congestion and pollution. This expansion would impose some hardships on those without automobiles and additional costs on the government for roads, services, and public transport. In rural areas, the effects will be more benign, bringing new opportunities for employment and other economic benefits with little added congestion or pollution. Local governments will have increased responsibilities for traffic management, regulation, and enforcement. The average cost of new cars will increase as national performance standards on emissions, efficiency, fuel quality, and safety are applied. As energy consumption almost certainly rises, China will become more dependent on imported petroleum.

Unfortunately, some of the important issues identified could not be explored in detail within the present study. For example, what effect will motorization have on inequities among various segments of the population? What will be the cost of and method of financing for the new infrastructure that will accompany increased motorization in China? To what extent will the air pollution from the emissions that will accompany the proliferation of cars in China have global implications for climate change? An expanding Chinese automobile fleet suggests that China will become a major petroleum importer; what effect would that have on the world demand for oil? Because these issues were beyond the resources available for the study, it was only possible to recognize their importance and to leave them for examination in later studies. Furthermore, the details and costs of implementation of the committee's recommendations will require a comprehensive analysis of specific urban development patterns, market forces, public revenues, and expenditures that was beyond the scope of this study.

THE EXPERIENCE OF OTHER COUNTRIES

Many other countries have gone through a similar motorization process, though few in so short a time. Some countries provide models for coping with the economic, environmental, and societal effects of rapid motorization in cities, and others offer successful examples of rapid development of a viable, independent automotive industry. Singapore, Hong Kong, Curitiba, and Houston provide different models for attempting to control urban traffic congestion, although none of them was trying simultaneously to develop a local automotive industry. The United States, Europe, and Japan have different approaches to controlling emissions and limiting fuel consumption. The development of an indigenous automotive industry can be observed in the varied experience of Japan and Korea, and a government-industry partnership to develop advanced vehicle technology was recently attempted in the United States. Each of these experiences offers a different lesson, and some were more successful than others in achieving their objectives. Though none of them is closely reproducible in the present situation, a careful examination of their experiences will be helpful to planners in China.

RESEARCH AND DEVELOPMENT

The Chinese automotive industry faces significant challenges in achieving independence. In the medium to long term, the industry must adjust to the effects of China's membership in the World Trade Organization (WTO), which will allow independent foreign automobile manufacturers and importers into China for the first time and may open the domestic market to an increased volume of imports that would challenge even the joint ventures in the growing Chinese market.

Presently, China spends a smaller percentage of its gross domestic product (GDP) on automotive research than any of the automobile-exporting countries and has limited trained and capable human resources in this area. Moreover, the current structure of its joint ventures with major producers does not encourage the transfer of technology to the Chinese partners. The result is that, at present, China does not have the independent capability to develop world-class vehicles. The Chinese automotive industry must, then, find ways to maintain currency with the research and development that is actively under way in several countries on the most promising technologies. In the short term, it will have to invest substantially to achieve capability in technologies such as advanced gasoline and diesel power trains, the application of sophisticated electronic controls, emissions control technologies, the use of new materials, and the application of complex engineering methods to optimize vehicle performance. To achieve long-term competitiveness in the world marketplace, the Chinese also must stay abreast of and contribute to emerging technical developments that may not materialize for 15–20 years. The challenge will be to maintain a proper balance between short-term and long-term needs. The industry cannot become competitive by concentrating its research only on one time frame. Meanwhile, it should attempt to develop partnerships with foreign producers that offer Chinese companies more access to evolving product and process technologies.

RECOMMENDED KEY ACTIONS

To meet the challenges presented by growing motorization and its environmental and social consequences, as well as China's entry into the WTO and the restructuring of its industry, the Chinese government and the automotive industry should consider the following key actions.

- The government should establish national standards for vehicle attributes such as low emissions, fuel efficiency, and safety. The goal set forth in the five-year plan for the automotive industry of adopting nationwide emissions standards on a par with European standards for vehicles and fuels by 2010 is an important step. Meanwhile, the Chinese government and automotive industry should develop and implement a process that will regularly assess the appropriate levels of vehicle performance standards, fuel standards, and infrastructure needs of the country.
- To achieve the 2010 emissions target, China must substantially improve the quality of fuels, with specific emphasis on drastically lowering the sulfur content. To meet European emissions standards, China also should adopt fuel quality standards identical to those of the European Union. Such a step will require major upgrades and the construction of new refineries, improved production efficiencies, and expanded foreign partnerships.
- State-owned motor vehicle enterprises should be restructured to

make them more competitive in the world marketplace. The government should facilitate the introduction of market-oriented management systems and the removal of social burdens.

• The government and the automotive industry must carefully consider their roles in developing the industry's capabilities. For example, they could jointly fund government-industry long-range R&D projects, the government could directly fund specific industry projects such as refinery upgrades and expansions, and industry could increase its spending on research and development. Research on advanced technologies, including hybrid vehicles, fuel cells, and alternative fuels, and on conventional technologies, including advanced gasoline and diesel power trains, pollution controls, electronic controls, and new materials, could provide the tools needed to design and develop world-class vehicles. Because foreign original equipment manufacturers (OEMs) are highly involved in the Chinese automotive industry, the government must soon decide how much these foreign members will be allowed to participate in any government-industry program.

• To meet the need for new technologies and attainment of environmental goals, the government should organize and support government laboratories and academic institutions in the pursuit of next-generation automotive technologies. Industry, including small companies, should be encouraged to participate. The government also should expand its support for academic programs that train students in automotive research and technology.

• Because the overseas joint venture partners possess much of the automotive technology that would be needed by an indigenous Chinese industry, ways should be sought to enhance the sharing of intellectual property. This may require that the joint ventures be restructured to allow the Chinese partners to participate more fully in research and development, and Chinese enterprises may need to limit their partnerships each to one foreign company to maximize trust and confidence, with appropriate long-term sharing of technology and knowledge.

• The highest growth sector in the future vehicle market is likely to be that for the small "China car" described in the five-year plan for the automotive industry. It should utilize technologies that allow high fuel efficiency, while providing reliable, comfortable, and safe transportation at an affordable cost. To be competitive, it should have attractive attributes that differentiate it from most imported cars in the growing Chinese market. Although a specific technical approach is not prescribed in the report, the various technical options for achieving both the near-term and the long-term product are described.

• The national, regional, and municipal governments must provide the infrastructural capacity needed for the increased motorization. In the cities, the municipal governments should undertake comprehensive development planning and provide additional road space and improve traffic management while minimizing social disruption. The new construction in cities should not be limited to ring roads and flyover highways; arterials also should be improved to relieve congestion in the neighborhoods and business areas while increasing the proportion of urban space dedicated to transportation. Meanwhile, the available schemes for road pricing should be explored to provide incentives for more efficient use of road space and financial resources for road maintenance and construction.

• As motorization proceeds rapidly in China, it is imperative that attention be directed to providing public transportation that is convenient, comfortable, sufficiently widespread, safe, and affordable. Recognizing that the fraction of the population that will own personal cars will be small for many years, China must maintain a balance between public transportation, nonmotorized vehicles, and private cars to ensure that the nondriving public, including bicyclists, is served adequately.

1

Introduction

The impact of the automotive industry on society is unlike that of any other industry. The automobile is not just a technology or mode of transportation; it is a fundamental determinant of the entire economy. In the United States, one of every six workers deals in some way with automobiles and trucks—making them, repairing them, driving them professionally, insuring them, licensing them, and building and maintaining highways for them. As for fuel, a steep rise in the price of gasoline seriously affects the entire U.S. economy, even causing a recession. Moreover, those nations that must import large quantities of oil find that their dependence has a telling effect on their balance of payments and creates potentially problematic relationships with the countries that produce oil, to the extent that dependence on foreign oil is an element of national security. Any nation's effort to build an automotive industry is, then, about far more than manufacturing cars and providing people with greater mobility. It is about changing the entire economic and employment structure of the nation.

China has one of the fastest-growing fleet of automobiles in the world. In 2001 its motor vehicles totaled 18 million, including 5 million cars, of which 600,000 were produced in 2000.[1] If China's number of motor ve-

[1] In this report the term *motor vehicles* excludes two-wheeled vehicles unless otherwise indicated. It does include cars, trucks, buses, and commercial vehicles. The 2001 figure for motor vehicles was obtained from the *China Statistical Abstract*, published by the China State Statistical Bureau in May 2002. The 2001 figure for cars is a projection based on 1998 data from the State Commission of Economy and Trade.

hicles per capita were comparable to the world average, its fleet would have to number 160 million, with 10 million new and replacement vehicles acquired each year. But, as revealed in Chapter 2, vehicle ownership is strongly correlated with a country's per capita gross domestic product (GDP) and average personal income. This suggests that a reasonable estimate for the number of vehicles in China in 2005 is about 25 million, a number in agreement with the general expectations of government planning.

CHINA'S FIVE-YEAR PLAN

In its eighth national five-year plan (1991–1995), the Chinese government designated the automotive industry as a "pillar industry" of the economy.[2] In its most recent five-year plan for the automotive industry (2001–2005), the government proposes specific actions to restructure and strengthen the industry, which is now primarily engaged in truck manufacture and in joint ventures with foreign manufacturers for automobile assembly, and to produce a Chinese family car at a price that would encourage mass ownership (for specifics of the five-year plan for the automotive industry, see the appendix to this chapter and Chapter 3). The plan emphasizes advanced technology and the production of vehicles that will be competitive in the international market, and gives priority to investments in highways and oil and gas pipelines. More specifically, the plan suggests that only two or three of the 118 existing automotive manufacturing companies will still be active in 2005, and it indicates which technologies will be used in manufacture and which will be incorporated into Chinese cars. Government planners anticipate that the new automobiles can be produced by independent Chinese companies, but they acknowledge that Chinese accession to the World Trade Organization (WTO) will complicate that prospect.

From the perspective of the Chinese automotive industry, the five-year plan represents short- and long-term opportunities and challenges. In the short term, the industry will receive government support to assist in consolidation and aid in expanding the domestic market. At present, nearly all the cars manufactured in China are produced by joint ventures, but the central government has indicated a desire to promote a Chinese-designed car, and additional resources may be made available for that effort. In the medium to long term, the automotive industry must adjust to the effects of China's membership in the WTO, which will open the

[2] The five-year plan is a national economic plan prepared by the central government of China. It is intended to provide guidance to both government and industry.

domestic market to a higher volume of imports that will challenge even the joint ventures in the growing Chinese market. In the longer term, Chinese companies will have to expand their capability in research and development (R&D), product design, and advanced manufacturing techniques in order to compete internationally in the car market. Such expansion will require major investments in research and development, which among successful international vehicle manufacturers is usually about 4–5 percent of revenues per year. To achieve that level of investment and support a high-quality research program, the China automotive industry would have to capture a significant share of the domestic market, in competition with joint ventures and imported vehicles, and at the same time conform to emissions and efficiency standards that will require the use of advanced technologies and high-quality fuels. There are, however, other kinds of vehicles, such as the agricultural vehicles and motorcycles in which China has a greater advantage and experience, that might be used to establish niches in the international market or to serve as a springboard to successful automobile manufacture and marketing, as they did for Honda (see Chapter 3).

Although implementation of the five-year plan will certainly bring about several improvements in the domestic motor vehicle industry and in the quality of products sold in the Chinese market, it is still not clear whether the Chinese industry will be able to nourish and sustain an R&D effort sufficient to develop an internationally competitive, indigenous vehicle. Success stories abound among China's neighbors in Asia, but China faces a difficulty they did not have. In addition to its need to compete with strong, mature industries in Japan, Europe, and the United States that are already well established in the region and in China itself, the Chinese government, as a member of the WTO, will find it difficult to protect its domestic industry as it enters the period of high growth. Indeed, the Chinese industry will face many challenges in seeking the capability to produce independently an internationally competitive product. Among other things, indigenous Chinese companies will have to expand their R&D capabilities, including in new product design and advanced manufacturing techniques, and they will have to invest heavily in manufacturing facilities so they can accommodate all new vehicle designs.

In any event, the growing automobile fleet, whether produced by indigenous Chinese industry, joint ventures, or imports, will present the Chinese people with both potential benefits and potential liabilities. In the short term, a more mobile population will have greater freedom of choice in housing location, employment, and leisure. But in urban areas, in the absence of government intervention, air quality will worsen, the number of automobile accidents will increase, and congestion will take a toll on the quality of life. In rural areas, increased motorization will likely produce less distressing effects and will bring new lifestyles and economic benefits without the immediate impact of congestion and pollution. In the longer term, the form of the cities will change, as they have nearly everywhere else in the world, and the population will move outward, leaving the city centers largely to the very wealthy and to the poorer people without automobiles. Although these effects can be mitigated, as they have been in some cities of the world, a new approach to government management and regulation will be required. Nationwide, the costs of car ownership will increase as national performance standards on emissions, efficiency, fuel quality, and safety are applied. Of great concern is the expectation that energy consumption will rise, and China will become more dependent on imported petroleum.

Many other countries have gone through a similar motorization process, though few in so short a time. Some countries provide models for coping with the economic, environmental, and societal effects of rapid motorization; others serve as successful examples of the rapid development of a viable, independent automotive industry. Singapore, Hong Kong (China), Curitiba (Brazil), and Houston (United States) provide different approaches for controlling congestion in cities, though they were not simultaneously trying to develop a local automotive industry. The United States and Europe have different approaches to controlling emissions. The United States, Europe, and Japan have active policies for limiting fuel consumption, but each has a different strategy. Japan, Korea, and Brazil have had varied experiences in the rapid development of an indigenous automotive industry, and the Partnership for a New Generation of Vehicles (PNGV) program in the United States has served as an example of a government-industry partnership to develop advanced technology. Each of these experiences, some more successful than others, offers a different lesson. None of these countries' experiences is closely reproducible in the present situation in China, but a careful examination of them will be helpful to Chinese planners.

INTERACADEMY COLLABORATION

In August 1999 a delegation from the Chinese Academy of Engineering (CAE) visited Washington, D.C., to enlist the collaboration of the U.S. National Academies in a joint study of the issues described in this chapter. When representatives of the National Academies visited Shanghai in October 1999, the institutions signed a memorandum of understanding and agreed on a plan to carry out a joint study (see Appendix A of this report).

The National Academies have a nearly 140-year history of providing the U.S. government and other governments and organizations with scientific and technological advice. Generally, this advice is tendered in the form of consensus study reports by balanced panels of experts, who are not paid for their contributions. The Chinese Academy of Engineering also is developing a program of reports and conferences that offer technical advice to the government.

For this study, the Chinese and U.S. academies formed a single committee of experts, half nominated by the CAE and half by the U.S. National Academies (the full list of committee members appears in the front matter). The committee was cochaired by Guo Konghui, a member of the Chinese Academy of Engineering, and W. Dale Compton, a member of the U.S. National Academy of Engineering. The committee met five times—in Beijing, Shanghai, and Changchun, China; Washington, D.C.; and Davis, California.

The purpose of the study was to:

- achieve an understanding of the benefits and costs of developing personal-use vehicles in China
- propose options for solving the problems of congestion, pollution, increased energy consumption, and changes in the urban structure
- suggest strategies for developing a Chinese national car, as described in the five-year plan, while taking into account China's social development, the available technologies, opportunities for international cooperation, the possibility of cooperation between government and industry, the role of such a car in the national transportation system, and the impact of a large increase in the number of private cars on sustainable development.

These issues are discussed in the chapters that follow. Chapter 2 describes the global patterns of motorization and compares China's present situation with that of other countries. Chapter 3 summarizes the present status of the Chinese automotive industry and describes some alternative future development paths. Conventional and advanced developments in automotive technologies that may be available in the near or long term to Chinese industry are described in Chapter 4. The status of the fuel and energy industries that will play a crucial role in the development of an environmentally and economically sustainable auto industry is reported in Chapter 5. Chapter 6 describes the probable societal effects of rapid motorization, including urban congestion, physical expansion of cities, and, at the same time, new lifestyle benefits for vehicle owners. The hazards associated with the atmospheric pollution caused by automobiles and the steps the Chinese government is taking to control emissions are described in Chapter 7. Chapter 8 is devoted to the prospects for government-industry cooperation for automotive research and development, and the last chapter presents the findings and recommendations of the joint study committee. The report also includes a case study of the effects of motorization on Shanghai (see Appendix B). It describes some of the municipality's innovative responses to congestion problems.

APPENDIX: CHINA'S FIVE-YEAR PLAN FOR THE AUTOMOTIVE INDUSTRY—ACHIEVEMENTS AND PROBLEMS

China's five-year plan for the automotive industry (2001–2005) identifies concrete goals and strategies for stimulating the rapid growth of the industry. (These goals and strategies are described in an appendix to Chapter 3.) Some of the achievements of the automotive industry and some of its remaining major problems also are listed in the plan. The achievements are:

- *High production advantages.* Eighty percent of the state's investment in the automotive sector is now concentrated in the 13 leading manufacturers.
- *Progress in product structure.* Cars as a proportion of total vehicle production rose from 8.3 percent in 1990 to 20.2 percent in 2000, and many of the new cars had electronic fuel injection systems.
- *Higher exports of parts and components.* Local suppliers provide up to 80 percent of the content of domestically produced vehicles, and they have been exporting 30 percent more parts and components each year.
- *Greater product development capability.* China has shifted from importing all manufacturing technology to developing some technology. New vehicle products have been developed jointly with foreign partners.
- *Significant progress in foreign economic and technology cooperation.* More than 600 automotive enterprises have been established in China with foreign participation. Foreign investment in the automotive sector has reached $21 billion (RMB174 billion).[3]

Four areas are listed as major problems:

1. *Uncultivated consumer market for automobiles.* The automotive sector has not developed a consumer-oriented policy—that is, the product line, marketing, and pricing of manufacturers are not tailored to meet consumers' needs. Moreover, some local governments impose fees and complex registration procedures or ban imports from other regions.
2. *Weak product development capability.* Chinese automobile manufac-

[3] Currency conversions are based on an exchange rate of U.S.$1 = RMB8.3.

turers lack a strong vehicle development capability, in part because of limited investment in research and development—less than 1 percent of revenue, which is much lower than that of most international manufacturers.

3. *Underdeveloped supply industry.* Despite the gains just noted, China's auto supply industry is fragmented and uncompetitive compared with international norms, with little economy of scale.

4. *Duplicate construction and serious fragmentation.* Local protective high tariffs and market restrictions have led local governments to launch new independent automobile assembly projects.

The plan also describes the effects of impending WTO membership in three areas:

1. *Tariff reduction and abolition of quota and import licenses.* Imports will place new pressures on domestic manufacturers, especially imports of passenger cars, advanced engines, drive axles and key parts assemblies, and high-end heavy-duty trucks.

2. *Opening up of the automobile service trade.* The arrival of foreign companies providing sales and distribution, franchise dealerships, shipping and transportation, financing, and car rental and leasing will further open channels for imports at the expense of locally produced vehicles.

3. *Abolition of the "localization policy."* The disappearance of China's present policy of preference for local enterprises will have a negative effect on foreign investment, technology transfer, and new product development in the Chinese automotive sector.

2

Motorization from a Global Perspective

China may benefit from the experience of other countries as it formulates its own policies for motor vehicle ownership and use, for the economic role of the motor vehicle industry, and for the use of new motor vehicle technology. To explore the potential for such learning, this chapter summarizes how the recent growth in motorization and road infrastructure in China compares with that of other countries. It also reviews policy issues related to motorization that may deserve special attention. Finally, the chapter explores recent work that predicts how changes in international trade practices and China's accession to the World Trade Organization (WTO) may affect the Chinese motor vehicle industry. Later chapters address motor vehicle technologies and the societal impacts of motorization.

GLOBAL COMPARISONS OF NATIONAL MOTORIZATION

Since the 1960s many studies have examined what determines the number of motor vehicles used in countries and cities over time.[1] All studies find that income is a major determinant of the size of the motor vehicle fleet across countries and cities in developing and industrial countries. At the

[1] See Ingram and Liu (1999) for a survey of this literature. In these studies motor vehicles are defined as those having four or more wheels and include cars, buses, trucks, and commercial vehicles, but not motorcycles. The statistics normally include government vehicles but usually not military vehicles. Income per capita is generally used in these studies and is usually measured using gross domestic product (GDP) per capita.

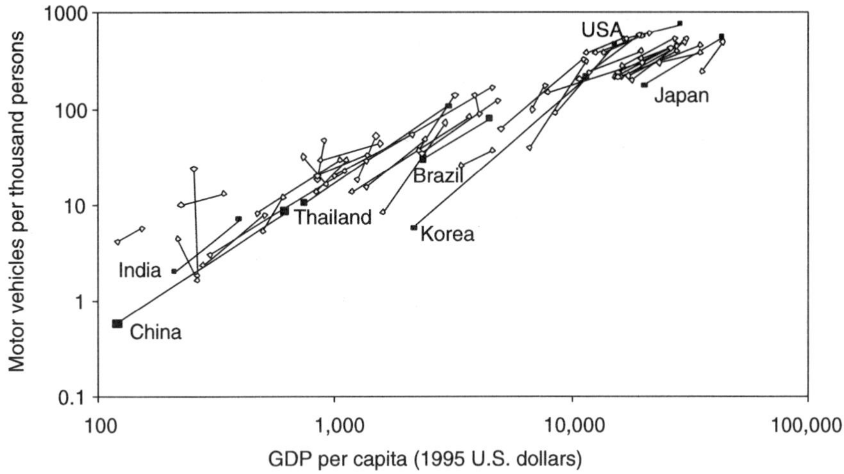

FIGURE 2-1 Motor vehicle fleets in relation to income, selected countries, 1970 and 1996. NOTE: Per capita gross domestic product (GDP) is transformed to dollars using market exchange rates (see footnote 2). SOURCES: Motorization data: International Road Federation (2001 and earlier); other data: World Bank (2001 and earlier).

national level, income alone typically explains more than 90 percent of the variation in motorization levels, and at the urban level more than 80 percent. The growth of national motor vehicle fleets parallels that of income: a 1 percent increase in income is associated with a 1 percent increase in motor vehicles, and this relationship has been relatively stable for the past 30 years.

The relation between motorization and income between 1970 and 1996 is summarized in Figure 2-1, which shows data for a sample of 50 countries, with seven countries identified.[2] Both per capita income and motorization levels vary over a nearly thousand-fold range, as shown by the logarithmic scales used. For each country in Figure 2-1 (and in Figures 2-2 and 2-3 later in this chapter) a line segment connects the country's position in 1970 with its position in 1996—the most recent year with comparable data across countries. By means of darker lines and end points, Figure 2-1 specifically identifies China and six other countries. Motorization increased in all countries from 1970 to 1996, but incomes did not. Downward sloping lines denote sampled countries (such as Nigeria, Rwanda, and Côte d'Ivoire) where incomes declined. A mostly parallel

[2] In all figures the GDP per capita, used to measure income or economic activity in this chapter, is transformed into dollars using market exchange rates because vehicles are traded goods, and the market exchange rate GDP measures the ability of an economy to purchase traded goods. The countries and data are in the appendix to this chapter.

alignment of the line segments indicates stability in the relation over time. The relation between motorization and income in China has been very consistent with that of other countries, even though China started from a very low income level in 1970.

Although strongly related to income, motorization can vary greatly—by a factor of two or more—across countries. For example, in 1970 Korea and Brazil had similar per capita incomes, but Brazil had 31.2 vehicles per thousand persons and Korea 5.6 vehicles per thousand persons. The United States, at 765 vehicles per thousand persons in 1996, had half again as many as some member countries of the Organisation for Economic Co-operation and Development (OECD) such as Germany at 529, France at 526, Japan at 546. Therefore, other variables, including nonmarket factors, also affect motorization levels.

Two additional factors significantly associated with national levels of motorization are population density (negatively) and urbanization (positively). No study has systematically examined the effect of domestic vehicle production on national motorization levels, but Korea's accelerated motorization may reflect a domestic industry effect. However, a look at two countries with similar income levels reveals that Thailand's motorization level is above Brazil's, yet Brazil has a much larger motor industry than Thailand (see Figure 2-1).

Vehicle and fuel prices are two policy instruments that often affect motor vehicle ownership and use. A comparison of current estimates shows that the elasticity of motorization with respect to income (about 1.0) is absolutely larger than the elasticity of motorization with respect to vehicle price (about –0.5). These magnitudes imply that if price increases were the only policy variable used to influence fleet size, vehicle prices would have to increase twice as fast as incomes to offset the effects of income growth on fleet size. Higher vehicle prices also increase the length of time the vehicle is retained (and thus the average age of the fleet). The effect of higher fuel prices on fleet size is uncertain.[3] However, it is known that an increase in fuel prices reduces vehicle usage and encourages the purchase of more fuel-efficient vehicles. When adjusted for inflation and quality, neither fuel nor vehicle prices have increased much over time, whereas income has grown steadily in most countries. Differences in prices and other factors across countries produce different levels of motorization at similar income levels (as shown in Figure 2-1), but income growth is the major determinant of motorization growth. China's experience is very consistent with this pattern.

---

[3] Compare Wheaton (1982), who finds a fuel price effect, with Johansson and Schipper (1997), who find none.

Studies of motorization at the urban level produce results generally similar to those at the national level. Income is strongly associated with urban vehicle fleet size, but motorization increases more rapidly with income at the national level than at the urban level because urban areas have substitutes for motor vehicles, such as transit, that rural areas lack. In addition, growth in urban motorization worsens urban congestion, which may make owning a car in a city less attractive. After controlling for income, one finds that motorization levels vary across cities even more than across countries (see Ingram and Liu, 1999:335).

THE AUTOMOBILE SHARE OF THE FLEET

Because this study focuses on personal-use motor vehicles (automobiles), it is now appropriate to ask: How does the automobile share of the motor vehicle fleet change as incomes rise over time? At low-income levels (less than $800 [RMB6,600] per capita), trucks and buses predominate. As incomes increase, the automobile share increases very rapidly at low income levels, then at a decreasing rate trending toward a saturation level that varies across countries. The relation between the automobile share of the fleet and income is shown in Figure 2-2 for the same 50 countries and

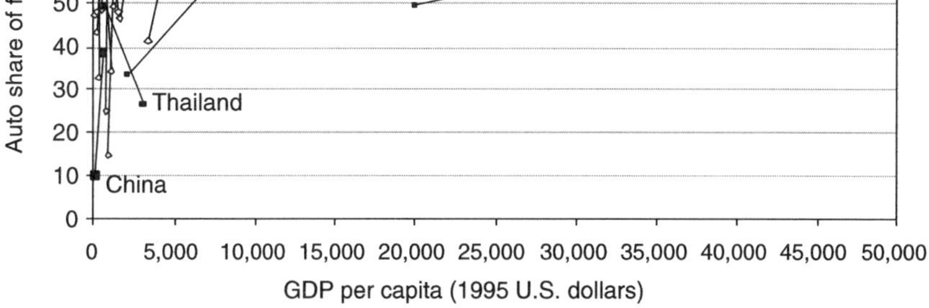

FIGURE 2-2 Automobile share of motor vehicle fleet in relation to income, selected countries, 1970 and 1996. NOTE: Per capita gross domestic product (GDP) is transformed to dollars using market exchange rates (see footnote 2). SOURCES: Motorization data: International Road Federation (2001 and earlier); other data: World Bank (2001 and earlier).

years, 1970 and 1996, used in Figure 2-1. The automobile share of the fleet has increased dramatically in China, Brazil, and Korea, whereas in the United States the growing popularity of pickup trucks, vans, and sport-utility vehicles has reduced the automobile share of the fleet since 1970. Surprisingly, the automobile share of the fleet also has fallen dramatically in Thailand, where pickup trucks have become popular for use in rural areas. And in some countries motorcycles are a popular substitute for cars. Thailand had 10.2 million motorcycles in 1996 (when China had 9.8 million) compared with 1.6 million automobiles and 4.6 million other motor vehicles.

The number of automobiles increases more rapidly with income than does the total number of motor vehicles. Thus when income rises by 1 percent, the number of motor vehicles rises by 1 percent, but the number of automobiles increases by about 1.2 percent. Compared with other countries, the share of the motor vehicle fleet composed of automobiles was at a very low level in 1970 in China (10.2 percent) and was still relatively low (38.9 percent) in 1996 in comparison with countries with similar income levels. But policies to limit the sales of motorcycles may increase automobile numbers.

GLOBAL COMPARISONS OF ROAD AVAILABILITY

Any policy adopted to increase the number of motor vehicles also needs to consider the availability of roads. Across countries, the size of the road network (measured as the length of paved and unpaved roads) increases less rapidly than income. However, as incomes rise more roads are paved. A 1 percent increase in income is associated with a 0.5 percent increase in total road length and a 1 percent increase in paved road length (see Figure 2-3). Again, China's recent experience is very consistent with that of other countries, and expansion of its network of paved roads has kept pace with its income growth. China's motor vehicle population, its paved road length, and its number of motor vehicles per kilometer of paved road (8.3 in 1996) are close to the average for its income level. In OECD countries the number of vehicles per kilometer (km) of paved road is higher, ranging from 20 to 60, with the higher numbers observed in countries with high population densities.

MOTORIZATION IN URBAN AREAS: A CHALLENGE

As noted earlier, in cities vehicle ownership increases with income, much as it does at the national level. Moreover, in developing countries urban incomes are often much higher than the national average income. In China, for example, average incomes in Shanghai are three to five times

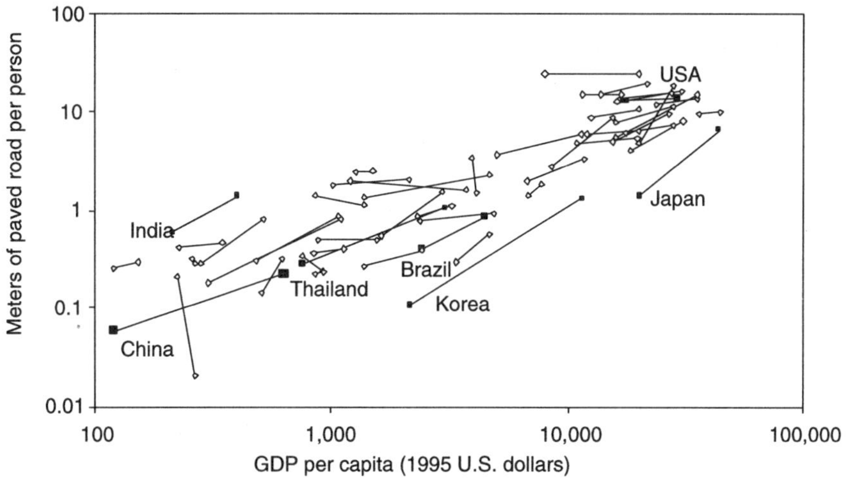

FIGURE 2-3 Paved road length in relation to income, selected countries, 1970 and 1996. NOTE: Per capita gross domestic product (GDP) is transformed to dollars using market exchange rates (see footnote 2). SOURCES: Motorization data: International Road Federation (2001 and earlier); other data: World Bank (2001 and earlier).

higher than the national average. Private car ownership would then be concentrated in Chinese cities. Unlike at the national level, the size of the road network at the urban level (measured as the length of roads within the urbanized area) increases very slowly with income. A recent analysis of 35 world cities found that a 1 percent increase in city per capita income was associated with only a 0.1 percent increase in urban road length—and most of that was from annexation of neighboring jurisdictions (Ingram and Liu, 1999). In China, road length has grown recently in major cities from both annexation and new road construction. But new urban roads are very costly. For example, in recent years Shanghai has invested 5 percent of its regional domestic product in roads. This share is high and is unlikely to be sustainable. Most countries invest about 1 percent of national income in roads per year (World Bank, 1994).

Road length per capita in large Chinese cities is similar to that observed in large cities in other developing countries. For example, Shanghai's road length of 0.43 m per person in 1997 was similar to that of Bangkok, Jakarta, and Manila. Because the number of motor vehicles has been increasing faster than the length of paved roads in large Chinese cities, the number of vehicles per kilometer of paved road has been rising. In 1997 Shanghai had about 122 vehicles per kilometer of paved road, a number that is high for cities in developing countries.

Urban income growth is associated with rapid growth in the number of vehicles and much slower growth in urban road length, and thus it worsens urban congestion—a negative externality of motorization. Because reducing urban congestion depends more on specific attributes of city land use and transport systems than of vehicles, most countries assign responsibility for congestion management to local or metropolitan governments. Such management includes traffic enforcement, parking control, and local charges for registration and vehicle use that may affect vehicle ownership. Chinese policy in this area is evolving. The central government recently restricted local government powers, abolishing many local fees on vehicle registration and use (China Online News, 2000b) and supporting the elimination of bridge tolls in Shanghai.

Vehicle-related air pollution, another negative externality of motorization, is primarily an urban problem because the density of vehicle use and population density are both high in cities. Although it may seem sensible to vary vehicle emissions standards by city, the mobility of vehicles and the cost to producers of multiple standards have led most countries to place responsibility for setting vehicle emissions standards at the national level. China recently assigned this responsibility to the State Environmental Protection Administration. Similar arguments support having vehicle safety standards set nationally, a practice China is also following.

Motorization in urban areas also has an impact on land use patterns. The growth in the use of trucks and motorized freight encourages firms to move out from the center of urban areas. The decentralization of employment encourages the suburbanization of residential development as workers follow jobs (Ingram, 1998). These locational changes reduce central city population densities and produce dispersed travel patterns that are less easily served by public transit. In cities, motor vehicle use promotes motor vehicle dependence because the residential changes and the housing and infrastructure investments are difficult to reverse. Nascent tendencies in this direction are evident in several large Chinese cities that are experiencing falling central population densities, growth in suburban employment, and the emergence of auto-dependent households in metropolitan suburbs.

OTHER MOTORIZATION POLICY LESSONS FROM MARKET ECONOMIES

Vehicle users pay many—but not all—of the costs of vehicle use in most countries. To ensure that personal vehicle use is not excessive, it is important that the prices paid by vehicle users reflect the full economic and social costs of vehicle use. When the personal vehicle user does not pay these appropriate costs, often some new regulation must be applied to, for example, vehicle attributes, safety, insurance, emissions, and vehicle use so that the owner is forced to pay the costs.

The vehicle owner normally pays the costs of personal vehicle ownership, operation, maintenance, and depreciation. Because vehicles use roads, it is important that vehicle users also pay an appropriate share of the cost of constructing and maintaining roads. Roads and related infrastructure are typically financed through taxes on fuel, taxes on use-related parts such as tires, and tolls. Road costs are usually about 5 percent of automobile operating costs.

Fuel costs, a quarter or more of operating costs, also are borne by the personal vehicle owner. Fuel prices affect both vehicle use and the fuel efficiency of the vehicle purchased. Taxes on fuels can be used to encourage car buyers to purchase more fuel-efficient vehicles. Such incentives may be desirable if vehicles use imported fuel and if the true cost to the economy of fuel imports is thought to be higher than the current market price. China, a net importer of petroleum since 1993, imported 70 million tons of oil in 2000, and oil imports are growing at 10 million tons per year. The Chinese government is considering auto fuel consumption policies that would incorporate a fuel tax and a tax on engine size to promote energy savings (China Online News, 2001b). Taxes on personal-use vehicles and vehicle fuel also may be an attractive source of general revenue for the government, because such taxes are normally progressive (revenue increases with the taxpayer's income).

Vehicles are involved in accidents, which damage property and injure people. Where vehicle insurance is required and vehicle operators are held liable for damage they cause, vehicle owners pay much of the cost of accidents through insurance premiums. In China, accident and motor vehicle-related fatality rates are high, although they are consistent with China's per capita income (Figure 2-4).[4] In 1996 the annual motor vehicle fatality rate per million vehicles was about 6,000 in China, which was about 30 times higher than the U.S. rate of 200. Auto insurance in China is growing rapidly, but it is unclear what insurance is mandatory and how liability for accidents is assessed. China's five-year plan for the development of the automotive industry describes the need for a road and motor vehicle law, but it does not mention insurance and liability for accidents. It is critical that users of personal vehicles be held responsible for the costs of the accidents they cause. A mandatory minimum level of

[4] Motor vehicle fatality rates are not available for many countries in the International Road Federation database. The data for China are derived from Chinese sources.

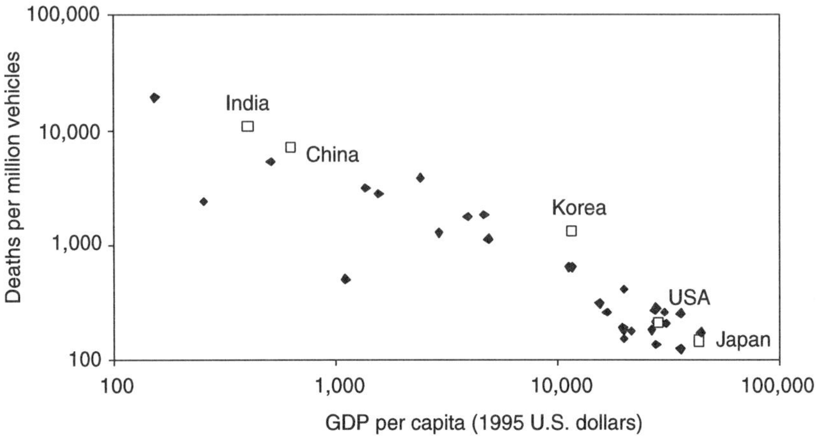

FIGURE 2-4 Motor vehicle death rates in relation to per capita income, selected countries, 1996. NOTE: Per capita gross domestic product (GDP) is transformed to dollars using market exchange rates (see footnote 2). SOURCES: Motorization data: Chinese sources (see footnote 4); other data: World Bank (2001 and earlier).

insurance is a sound way to ensure that vehicle users pay the costs of accidents and the damage associated with vehicle use.

Emissions from motor vehicles often harm urban air quality, but one vehicle's emissions generally cause no costs to its operator. Nor do vehicle operators pay for the health care costs of those who suffer from the resulting poor air quality. This negative externality is generally handled by regulations and controls on engine technology and fuel use rather than by prices or taxes. China has moved to adopt international standards in regulating vehicular emissions. Fuels are regulated—for example, leaded gasoline has been banned—and new emissions standards promulgated by the State Environmental Protection Administration require light-duty vehicles in China to meet European Emission Standard II (Euro II) by the year 2004 (China Online News, 2001a). Some metropolitan areas have inspection and maintenance programs for vehicular emissions systems. But nationally mandated inspection standards for emissions and safety have been put in place and are being strengthened.

The use of motor vehicles in urban areas also imposes other costs that vehicle operators do not pay. Vehicle use on crowded streets increases the travel time of all other vehicles, but vehicle operators incur only the cost of the time they spend on their trip. Economists have long advocated congestion tolls for crowded roads, and Singapore has successfully implemented one. Implementing congestion tolls can be difficult, but intelligent transportation systems (ITS) could be used successfully to levy congestion tolls.[5] Traffic management remains the second-best solution to congestion and is better left in the hands of local officials.

PROJECTIONS OF CHINA'S MOTOR VEHICLE FLEET

The strong relation between income and motorization across countries provides a simple basis for projecting motor vehicle fleet sizes in China, because China's motorization experience has been so consistent with international patterns.[6] Cross-country analysis has indicated that motor vehicle fleets grow in proportion to income and that automobile fleets grow 1.2 times faster than income. Those relationships are used here, but projecting the growth rate of the Chinese economy is a challenge. China's gross domestic product (GDP), an indicator of personal income, grew at an average annual rate of 10.1 percent from 1980 to 1990 and 10.7 percent from 1990 through 1998 (World Bank, 2001:194).[7] However, in 1999 and 2000 it grew at 7.1 and 8.0 percent, respectively (World Bank, 2001:192). Decadal growth rates of 10 percent or more are rare in historical experience.

Three different assumptions about China's GDP growth rate—a high rate of 10 percent, a medium rate of 8 percent, and a low rate of 6 percent—produce three different projections for the automobile and motor vehicle fleet size in China. (Note that very few countries have attained even the 6 percent rate over a decade.) The projections, starting from a base in 1996 of 10,020,000 motor vehicles (including 3,894,000 automobiles), are shown in Table 2-1. They cover a wide range, differing by a factor of two by 2015. The medium projection is close to the projections made recently by the Chinese Academy of Engineering.[8]

It is difficult to move from projections of fleet size to projections of annual new vehicle sales. Taxes on fuel mainly affect fuel efficiency rather than the number of vehicles sold. High prices or high taxes on new vehicles tend to increase the economic life of vehicles. Longer vehicle life mean a smaller share of the fleet is scrapped each year, and thus new

---

[5] Intelligent transportation systems are produced by using advanced technologies—such as computers, electronics, and communications—to integrate surface transportation systems with the goal of improving safety and efficiency.

[6] This consistency is not automatic and will occur only if policies allow vehicle growth to continue.

[7] Projections of motor vehicle fleet size are made using total income and not income per capita, because empirical estimates indicate that population is a scale variable (has an exponent of 1.00) in regressions of total fleet size on per capita income and population.

[8] These projections were presented by Prof. Guo Konghui in May 2001, in Beijing, China.

TABLE 2-1 National Vehicle Fleet Projections for Three GDP Growth Rates, China (millions of vehicles)

Year	Cars	Motor Vehicles
10 percent GDP growth		
2005	7.9	26.4
2010	13.9	42.5
2015	24.4	68.4
2020	43.1	110.2
8 percent GDP growth		
2005	7.2	24.5
2010	11.4	36.0
2015	18.0	52.9
2020	28.5	77.8
6 percent GDP growth		
2005	6.6	22.7
2010	9.3	30.4
2015	13.2	40.7
2020	18.7	54.5

NOTE: Projections (in millions of vehicles) assume that motor vehicle growth is the same as income growth and that car growth is 1.2 times income growth. GDP = gross domestic product.

SOURCE: Calculated by G. Ingram based on data in China State Statistical Bureau (2002).

vehicle sales are a lower share of the fleet. Chinese policy makers recently have set rules calling for vehicles to be retired at a certain mileage and age, but these rules seem to reflect safety and emissions standard objectives rather than economic considerations (China Online News, 2000a).[9]

CHANGES IN INTERNATIONAL TRADE POLICIES

In 1999 China was the seventh largest national economy as measured by GDP valued at market exchange rates and the second largest (behind the United States) when GDP is measured using purchasing power parity

[9] These standards call for retiring vehicles that have reached 30,000–50,000 km or eight to ten years of age.

BOX 2-1
Main Obligations of WTO Members

Members of the World Trade Organization are obligated to:

- Offer the same trade policies (most-favored-nation status) to all WTO members.
- Make government legislation and domestic enforcement conform to WTO standards.
- Reduce tariffs and rely only on tariffs; remove quotas, most licenses, and other nontariff barriers.
- Subscribe to:
    - *General Agreement on Trade and Tariffs (GATT)*—schedule of tariffs countries have bound themselves to follow.
    - *General Agreement on Trade in Services (GATS)*—agreement modeled on the GATT and its principles and that applies to national, state, and local governments.
    - *Trade-Related Aspects of Intellectual Property Rights (TRIPS)*—agreement that protects copyrights, trademarks, geographical indications, industrial design, patents, integrated circuit design, and trade secrets.
    - *Technical Barrier to Trade (TBT) Agreement*—agreement that controls the use of technical regulations and standards as trade barriers.
    - *Agreement on Trade-Related Investment Measures (TRIMs)*—agreement that specifies that investors cannot be required to have local content minima or to balance exports and imports.
    - *Antidumping provisions*—provisions that prohibit selling goods below cost if local industry is injured.
    - *Sanitary and phytosanitary (SPS) measures*—measures that ensure that health regulations are not discriminatory.

SOURCE: World Trade Organization Secretariat (1999).

exchange rates. In international trade flows, China ranked fourth behind the United States, the European Union, and Japan. China was accepted into the World Trade Organization in late 2001; it was not a member of the WTO's predecessor, the General Agreement on Tariffs and Trade (GATT).[10] China steadily liberalized its trade policies during the 1990s, but joining the WTO will involve many additional changes and obligations (see Box 2-1).

[10] The GATT is no longer a free-standing body; to join the GATT a country must join the WTO.

The Automotive Industry and Trade Liberalization

China's accession to the WTO will have a large impact on its automotive industry, an economic sector that is heavily involved in international trade. For example, in 1999 machinery and transport equipment constituted 49 percent of the value of exports (and 30 percent of the value of imports) of the OECD's trade with low- and middle-income countries (World Bank, 2001:327–328). In 1999, of the new cars placed in service in China, 95 percent were produced domestically.

China's automotive industry is relatively fragmented; as of 1999 it had 118 original equipment manufacturers of motor vehicles. Its overall production capacity for cars was 910,000 at the end of 1999, and annual production was 605,000. The three largest companies—First Auto Works (FAW), Dongfeng Motor Corporation (DMC), and Shanghai Automotive Industry Corporation (SAIC)—produced 44 percent of motor vehicles and 70 percent of the cars during this period. In 2000 the 13 largest automotive companies produced over 90 percent of the total motor vehicle output and sales.

The automotive industry is one of China's most highly protected. Tariffs on motor vehicles and vehicle parts were well over 100 percent as recently as 1995. In 2000 import tariffs on sedans ranged from 80 to 100 percent, and import tariffs on vehicle parts ranged from 35 to 50 percent. After WTO accession, compliance with WTO standards must be phased in over a period of five years, and average motor vehicle tariff rates will fall below 15 percent—with tariffs on vehicles at 25 percent and on components around 10 percent.[11] The import quotas and import licenses applied to all vehicles except cars will be removed in the second half of the five-year phase-in period and those for cars at its end. Although the tariff rates are high, eliminating quotas and licenses may have more of an impact. Indeed, the five-year plan for the automotive industry states that the elimination of import quotas will be much greater than that of reduced tariffs. It also notes that WTO accession will most severely affect cars, followed by heavy-duty trucks; it will have minimal effects on mini-vehicles, medium trucks, buses, and motorcycles (China State Economic and Trade Commission, 2001:6).

The automotive industry in China has many small producers, and even its largest companies have few plants that operate on a scale large enough to achieve least-cost output levels. High local content has then been achieved, but at a high cost. These challenges are well recognized by China's economic policy makers and are addressed by current policy

[11] Many electronic components of vehicles may come in with a zero tariff under the Information Technology Agreement.

statements. China's objectives over the next five years are to restructure its automotive industry, to consolidate its many small-scale producers into three large companies with a joint domestic market share of over 70 percent, and to establish the capacity to develop vehicles domestically by increasing expenditures on research and development.

The Effects of WTO Accession on China's Automotive Industry

A few scholars have attempted to develop quantitative estimates of the economic effects of China's accession to the WTO (Development Research Centre, 1998; Fan and Zheng, 2000; Zhai and Li, 2000; Wang 2001). One recent paper developed estimates for 22 separate industries, including the automotive industry, using a computable general equilibrium model that includes China and other countries or regions of the world (Ianchovichina and Martin, 2001). Its results are briefly summarized here.

Ianchovichina and Martin conducted simulations for 1995–2005 that compared China's accession to the WTO with a non-WTO base case. Their model incorporates the liberal duty exemptions already in place for imports used as inputs into export goods and for investment goods used in joint ventures with foreign enterprises. Ianchovichina and Martin report that, overall, WTO accession produces substantial benefits for China and the rest of the world. China's overall share of world exports is projected to double, with labor-intensive industries growing the most—especially exports of apparel.

In the 1995 base year, of the 22 industries included in the simulation the Chinese automotive industry had the highest tariffs, at 129 percent, compared with the average weighted tariff across all industries in 1995 of 21 percent. With China's accession to the WTO, import tariffs on automotive trade fall to 14 percent (an average of vehicle and component tariffs), compared with an industry-wide average weighted tariff of 7.85 percent. The model predicts for China a substantial increase in automotive exports and an even larger increase in automotive imports. The net result is a prediction that the total output of China's automotive industry will decline slightly (the only industry to do so) by 2005 under WTO accession. The authors are careful to note that their model does not capture economies of scale in automobile production. If reduced protection forces consolidation and increased economies of scale, China's automobile sector could become a stronger competitor and a larger exporter. Although conclusions from these simulations are subject to large uncertainties, they do indicate that WTO accession will place tremendous strains on the automotive industry in China, because it has been highly protected and has few producers with enough volume to benefit from scale economies.

Outside of China, the six large global auto companies that have emerged from mergers during the 1990s each have the capacity to produce about 4 million units per year and are well poised to compete for a market share in China after WTO accession. In China, vehicle imports are likely to grow rapidly after WTO accession. Imported vehicles and auto parts will put competitive pressure on existing domestic producers and will likely lead to a reduction in the number of vehicle producers and the elimination of inefficient smaller firms. Exports of auto parts are likely to increase as both Chinese and international firms take advantage of China's relatively low wage levels.

CONCLUSION

In contrast to the impressions of casual observers, China's pattern of motorization is thus far very similar to that of other countries, and its overall motor vehicle fleet size is strongly associated with its growing income level. Current policies are likely to sustain this similarity. Likewise, the pattern of paved road length in China is similar to that of other countries.

One issue common to many other countries—and rapidly emerging in China—is that motorization is likely to produce its earliest and severest problems in cities. Urban residents have higher-than-average incomes and therefore buy automobiles earlier and at a higher rate than the general population. Because urban vehicle fleets grow more rapidly than urban road length, urban congestion increases quickly—promoting the decentralization of population and employment from central to peripheral locations. Such patterns have already been observed in China's largest cities.

Overall, motorization produces negative externalities that must be regulated. In many areas, China has put in place regulations (on emissions, fuel quality, and crash-worthiness) that are quite advanced and draw on the experience of other countries. To date, China's nationally mandated vehicle inspection programs seem to focus mainly on vehicle safety and to apply near the expected end of a vehicle's useful life. China may well wish to ensure that motor vehicle users pay the costs of accidents through a mandatory insurance program or a financial responsibility requirement.

China's accession to the World Trade Organization will result in lower tariffs by a factor of two to four from the present levels on vehicles and components. In addition, import quotas and import licenses will be eliminated. These changes, to be phased in over a five-year period, will bring competitive pressure to bear on China's automotive industry, which is currently fragmented and enjoys few scale economies. Some detailed quantitative analyses suggest that the output of China's automobile industry will be essentially unchanged by 2005, and that both imports and exports of motor vehicle-related goods will grow. The implication is that much of the automobile growth over the five-year period will stem from imports. Although subject to considerable quantitative uncertainty, these analyses signal that China's automotive industry may face a significant structural adjustment from WTO accession.

APPENDIX 1
NATIONAL DATA

National Data, 50 Countries, 1970 and 1996

Country	1970 GDP per capita (1995 US$)	1996 GDP per capita (1995 US$)	1970 Land area (km^2)
Algeria	1,261	1,503	2,381,740
Argentina	6,833	7,743	2,736,690
Australia	13,636	21,347	7,682,300
Austria	16,053	29,813	82,710
Belgium	15,736	27,415	32,820
Bolivia	856	920	1,084,380
Brazil	2,393	4,476	8,456,510
Cameroon	508	617	465,400
Canada	12,460	19,820	9,220,970
Chile	2,360	4,858	748,800
China	120	630	9,327,450
Colombia	1,377	2,403	1,038,00
Cote d1voire	927	747	318,000
Denmark	23,446	35,115	429370
Ecuador	879	1,564	276,840
Egypt	478	1,066	995,450
Finland	15,200	26,239	304,590
France	16,412	27,060	550,100
Gabon	3,390	4,634	257,670
Germany	17,988	30,237	349,270
Greece	6,651	11,488	128,900
India	212	400	2,973,190
Indonesia	298	1,105	1,811,570
Ireland	7,908	19,685	68,890
Italy	10,801	19,331	2949090
Japan	20,015	42,913	376,520
Kenya	226	342	569,140
South Korea	2,641	11,467	98,730
Malawi	121	153	94,080
Malaysia	1,371	4,625	328,955
Mauritius	1,190	3,703	2,030
Mexico	2,295	3,251	1,908,690
Morocco	849	1,379	446,340
Netherlands	17,321	27,544	33,780
New Zealand	12,685	16,588	267,990
Nigeria	264	256	910,770
Norway	15,669	35,102	306,830
Pakistan	277	512	770,880
Philippines	845	1,122	298,260
Portugal	5,016	11,203	91,500

1996 Land Area (km^2)	1970 Total Population	1996 Total Population	1996 Motor Vehicle-Related Deaths
2,381,740	13,700,000	28,600,000	
2,736,690	24,000,000	35,200,000	
7,682,300	12,500,000	18,300,000	1,970
82,730	7,426,000	8,059,000	1,027
32,820	9,638,000	10,200,000	1,356
1,084,380	4,212,000	7,588,000	
8,456,510	96,000,000	162,000,000	
465,400	6,614,000	13,500,000	
9,220,970	21,300,000	29,700,000	3,082
748,800	9,496,000	14,400,000	1,925
9,327,420	818,000,000	1,220,000,000	69,000
1,038,700	22,600,00	39,300,000	7,445
318,000	5,515,000	14,300,000	
42,430	4,9299000	5,262,000	514
276,840	5,970,000	11,700,000	1,421
995,450	33,100,000	59,300,000	
304,590	4,606,000	5,125,000	404
550,100	50,800,000	58,000,000	8,080
257,670	504,000	1,125,000	
349,270	77,700,000	81,900,000	8,758
128,900	8,793,000	10,500,000	2,068
2,973,190	548,000,000	946,000,000	71,943
1,811,570	118,000,000	197,000,000	
68,890	2,950,000	3,632,000	453
294,060	53,800,000	579400,000	6,193
376,520	104,000,000	126,000,000	9,942
569,140	11,500,000	279500,000	
98,730	31,900,000	45,500,000	12,653
94,080	4,518,000	10,000,000	1,090
328,550	10,900,000	21,100,000	6,304
2,030	826,000	19134,000	
1,908,690	50,600,000	92,600,000	
446,300	15,300,000	26,800,000	2,807
33,920	13,000,000	15,500,000	1,334
267,990	2,820,000	3,714,100	514
910,770	53,200,000	114,000,000	6,364
306,830	3,877,000	4,381,000	255
770,880	60,600,000	125,000,000	5,280
298,170	37,500,000	69,900,000	19043
91,500	9,044,200	9,930,000	2,100

Continued on next page

National Data, 50 Countries, 1970 and 1996 (continued)

Country	1970 GDP per capita (1995 US$)	1996 GDP per capita (1995 US$)	1970 Land Area (km^2)
Rwanda	263	221	24,670
South Africa	4,100	3,943	1,221,040
Spain	8,507	15,224	499,780
Sweden	19,598	27,454	411,620
Switzerland	35,491	43,574	39,550
Thailand	752	3,021	510,890
Tunisia	1,004	2,119	155,360
Turkey	1,626	2,943	769,630
United Kingdom	11,827	19,651	241,770
United States	17,052	28,341	9,159,120

Country	1970 Total Motor Vehicles ('000)	1996 Total Motor Vehicles ('000)	1970 Passenger Cars ('000)
Algeria	251	1,505	143
Argentina	2,318	6,071	1,440
Australia	4,870	11,097	3,899
Austria	1,575	3,994	1,197
Belgium	2,302	4,768	2,060
Bolivia	77	362	19
Brazil	3,000	12,754	1,595
Cameroon	35	162	20
Canada	8,340	16,861	6,602
Chile	328	1,720	176
China	488	10,020	50
Colombia	343	1,922	239
Cote d'lvoire	89	456	56
Denmark	1,471	2,026	1,079
Ecuador	180	518	27
Egypt	270	1,787	131
Finland	997	2,210	712
France	14,370	30,558	12,900
Gabon	13	41	6
Germany	15,663	43,351	14,673
Greece	344	3,246	227
India	1,092	6,684	627
Indonesia	359	4,439	239

1996 Land Area (km^2)	1970 Total Population	1996 Total Population	1996 Motor Vehicle-Related Deaths
24,670	3,728,000	6,727,000	
1,221,040	22,100,000	39,900,000	9,848
499,40	33,800,000	39,300,000	5,483
411,620	8,043,000	8,843,000	537
39,550	6,267,000	7,074,000	616
510,890	35,700,000	59,000,000	
155,360	5,127,000	9,089,300	
769,630	35,300,000	61,400,000	5,428
241,600	55,600,000	58,800,000	3,598
9,159,120	205,000,000	268,000,000	42,065

1996 Passanger Cars ('000)	1970 Road Length ('000 km)	1996 Road Length ('000 km)	1970 Paved Roads ('000 km)	1996 Paved Roads ('000 km)
725	76.0	104.0	33.0	71.6
4,784	201.1	218.2	33.4	63.5
9,022	884.7	913.0	185.8	353.0
3,691	94.8	129.1	94.8	129.1
4,308	92.1	144.1	75.1	116.1
224	25.6	41.6	0.9	1.8
10,500	1,130.0	1,670.1	38.9	139.7
98	46.6	70.1	0.9	4.3
13,217	830.3	901.9	186.9	318.4
1,121	64.5	79.1	7.4	13.2
3,894	636.7	1,210.0	47.0	271.0
1,624	49.5	129.1	6.0	15.4
293	35.0	50.4	1.3	4.9
1,737	61.5	71.3	57.6	71.3
465	20.6	43.2	2.9	5.7
1,354	23.6	64.0	10.1	50.0
1,943	72.4	77.8	23.2	49.8
25,500	785.2	892.5	691.0	892.5
25	6.0	8.3	0.2	0.6
40,988	440.9	656.1	317.4	650.0
2,339	35.1	38.1	17.4	34.3
4,189	972.3	2,367.0	324.8	1,295.0
2,409	84.3	336.4	21.1	155.8

Continued on next page

National Data, 50 Countries, 1970 and 1996 (continued)

Country	1970 Total Motor Vehicles ('000)	1996 Total Motor Vehicles ('000)	1970 Passenger Cars ('000)
Ireland	440	1,109	394
Italy	11,115	33,316	10,181
Japan	17,826	68,805	8,832
Kenya	114	359	96
South Korea	180	9,553	61
Malawi	18	56	9
Malaysia	312	3,497	238
Mauritius	11	94	6
Mexico	1,825	12,818	1,234
Morocco	306	897	223
Netherlands	2,913	6,260	2,258
New Zealand	1,080	1,987	891
Nigeria	98	2,701	57
Norway	835	2,053	694
Pakistan	146	977	93
Philippines	510	2,053	279
Portugal	553	3,263	510
Rwanda	6	30	3
South Africa	1,973	5,657	1,545
Spain	3,125	17,860	2,378
Sweden	2,690	3,981	2,289
Switzerland	1,524	3,546	1,383
Thailand	376	6,234	185
Tunisia	104	494	66
Turkey	298	4,328	138
United Kingdom	13,330	23,392	11,666
United States	109,305	205,146	88,840

SOURCES: Per capita gross domestic product (GDP), land area (square kilometers), and population: World Bank's World Development Indicators Database. Motor vehicle data,

1996 Passanger Cars ('000)	1970 Road Length ('000 km)	1996 Road Length ('000 km)	1970 Paved Roads ('000 km)	1996 Paved Roads ('000 km)
987	86.7	92.5	71.6	86.9
30,600	285.0	303.9	262.2	303.9
46,869	1,013.6	1,147.5	152.0	825.6
278	41.5	67.2	4.8	12.5
6,894	40.2	82.3	3.6	59.8
27	10.7	14.6	1.1	2.9
2,946	22.6	63.4	14.8	47.2
70	1.8	1.9	1.6	1.8
8,707	72.3	312.3	42.7	99.3
670	45.9	60.7	21.1	30.5
5,664	79.9	124.1	78.6	112.0
1,636	93.8	91.9	41.9	55.9
885	89.0	112.9	15.2	36.3
19661	72.3	91.3	21.7	65.7
578	31.7	224.9	17.5	98.9
703	75.7	161.3	13.5	28.1
2,671	41.8	68.7	32.4	59.1
13	6.5	14.9	0.1	1.4
4,004	185.5	331.3	33.1	137.5
14,754	139.4	344.8	94.7	341.0
3,655	98.0	210.0	38.6	162.0
3,268	59.2	71.1	59.2	71.1
1,661	16.3	200.3	10.0	62.9
320	17.9	29.2	9.1	18.2
3,457	59.5	381.6	19.0	95.4
21,172	334.1	368.8	324.2	368.8
129,728	6,003.0	6,308.1	2,668.9	3,816.4

including deaths: International Road Federation (IRF, various years). (Motor vehicle fatality rates are not available for many countries in the IRF database.) Length of total roads and paved roads: IRF (various years); World Bank (1994); Canning (1998).

REFERENCES

China Online News. 2000a. Standards for retiring vehicles published. August 29. Online. Available at www.chinaonline.com. Accessed October 21, 2002.

———. 2000b. Taking the freer way: China eliminates 238 vehicle fees to encourage automobile ownership. June 6. Online. Available at www.chinaonline.com. Accessed October 21, 2002.

———. 2001a. China unifies auto-emission standards. April 23. Online. Available at www.chinaonline.com. Accessed October 21, 2002.

———. 2001b. New policies encourage purchase of small engine sedans. January 23. Online. Available at www.chinaonline.com. Accessed October 21, 2002.

China State Economic and Trade Commission. 2001. China's 10th Five-Year Plan for the Development of the Automotive Industry (2001–2005). June 26. (Translation made available by Volkswagen Corporation.)

China State Statistical Bureau. 2002. China Statistical Abstract. May.

Development Research Centre. 1998. The Global and Domestic Impact of China Joining the World Trade Organization. Development Research Centre of State Council, Beijing.

Eglin, R. 2000. Challenges and Implications of China Joining the WTO. June 19. World Trade Organization Secretariat, Geneva, Switzerland.

Fan M. and Zheng Y. 2000. China's trade liberalization for WTO accession and its effects on China: A computable general equilibrium analysis. Paper presented at Third Annual Conference on Global Economic Analysis, Monash University, June 28–30.

Ianchovichina, E., and W. Martin. 2001. Trade liberalization in China's accession to the World Trade Organization. Journal of Economic Integration 16(December):421–445.

Ingram, G. K. 1998. Patterns of metropolitan development: What have we learned? Urban Studies 35(7):1019–1035.

Ingram, G. K., and Liu Z. 1999. Determinants of motorization and road provision. Pp. 325–356 in Essays in Transportation Economics and Policy, J. Gómez-Ibánez, W. B. Tye, and C. Winston, eds. Washington, D.C.: Brookings Institution Press.

International Road Federation. 2001. World Road Statistics. Geneva.

Johansson, O., and Schipper, L. 1997. Measuring the long-run fuel demand of cars: Separate estimations of vehicle stock, mean fuel intensity, and mean annual driving distance. Journal of Transport Economics and Policy 31(September):277–292.

Wang Z. 2001. The impact of China's WTO accession on patterns of world trade. Paper presented at International Agricultural Trade Research Consortium, May 18–20.

Wheaton, W. C. 1982. The long run structure of transportation and gasoline demand. Bell Journal of Economics 13(autumn):439–454.

World Bank. 1994. World Development Report, 1994: Infrastructure for Development. Washington, D.C.

———. 2001. World Development Indicators. Washington, D.C.

World Trade Organization Secretariat. 1999. Guide to the Uruguay Round Agreements. London: Kluwer Law International.

Zhai F. and Li S. 2000. The implications of accession to WTO on China's economy. Paper presented at Third Annual Conference on Global Economic Analysis, Monash University, June 28–30.

3

Structure and Capability of China's Automotive Industry

The Chinese government is seeking to develop an automotive industry that is fully competitive with the world's leading original equipment manufacturers (OEMs)—an ambitious goal, with many implications for society and for the Chinese economy. This chapter identifies the challenges and barriers that must be overcome if the Chinese automotive industry is to achieve this goal and the level of independence from foreign technology called for in the most recent five-year plan for the automotive industry (2001–2005). In doing so, the chapter draws on the experience of automotive companies throughout the world and looks at how the governments of a variety of countries have helped their automotive industries.

EVOLUTION OF THE CHINESE AUTOMOTIVE INDUSTRY

China began to develop a domestic motor vehicle industry in the 1950s. By pooling together investment and imported technology, primarily from the Soviet Union, the government was able to undertake establishment of the First Auto Works (FAW) in Changchun in 1953. On July 15, 1956, the first Chinese-made vehicle was produced—a 4-ton truck.

By 1958 many local governments were investing in the automotive industry, with the result that more than 200 factories began to produce motor vehicles. Yet only a small number of these factories survived and went on to become the backbone of today's automotive industry. Those that did were in Beijing, Nanjing, Shanghai, Shenyang, and Jinan. One product of these plants was the Red Flag sedan, the limousine used by high-ranking leaders in China. By 1960 the annual output of motor vehicles exceeded 22,000, but the industry then went into a decline, producing less than 4,000 vehicles in 1961, and the original production scale was not resumed until 1963.

In the late 1960s China began to build the Second Auto Works, which later became the Dongfeng Motor Corporation (DMC). It was located in a valley in northwestern Hubei Province, a mountainous region. The Second Auto Works reached its designed production capability in 1986 and began to produce 5-ton trucks. Other heavy-duty truck manufacturers, such as the Sichuan Auto Plant and the Shannxi Auto Plant, also appeared during this period. They too were built in mountainous areas, which impeded production and further development.

In 1971, after a decade of development, the total output of China's automotive industry exceeded 100,000 units. Growth remained slow, however, with the total annual output still under 150,000 seven years later. In the 1970s the total number of motor vehicle manufacturing facilities increased to over 50, but most of them were small and had low production.

Earlier, in the 1960s, the Chinese government had attempted to implement a highly centralized management system for the automotive industry, but for many years the industry developed in a scattered and disorderly fashion. With economic reform in the 1980s, the highly centralized control of motor vehicle production under the planned economy was gradually replaced by a market-oriented approach. The product mix was adjusted, and the production of heavy-duty and light-duty vehicles[1] was expanded to eliminate the shortage of these vehicles. China also stepped up its cooperation with automotive industries in other countries, importing technology and establishing joint ventures. In May 1983 Beijing Jeep Corporation, the first joint venture for manufacturing complete vehicles, was established. Later, Shanghai-Volkswagen, FAW-Volkswagen, Dongfeng-Citroën Company, and other joint ventures came into being. Adjustments also were made in the structure of the industry, and a group production and management system was gradually created. During the 1980s annual motor vehicle output increased rapidly, from slightly more than 200,000 in 1980 to almost 600,000 in 1989.

During the 1990s China's automotive industry further adjusted its strategy, placing much higher priority on the development of the passenger car industry. Before the 1980s China did not allow private citizens to purchase motor vehicles for personal use and therefore did not develop passenger car production. In the mid-1980s, when the control on private pur-

---

[1] According to the National Standard of the Peoples Republic of China GB 3730.1-88, a medium-duty truck weighs between 6 and 14 tons. Light-duty trucks are lighter, and heavy-duty trucks heavier.

chase was lifted, the number of personal-use automobiles began to grow. By 1985, 3,500 cars a year were being imported. Thus the development of car production to meet a growing demand became a government priority, and between 1990 and 2000 annual car production grew from less than 50,000 to over 600,000 at plants in Shanghai, Beijing, Tianjin, Guangzhou, Chongqing, and Guizhou, including those of the FAW Group and DMC.

After lagging behind that of the industrialized world for many years, China's automotive industry, and especially the personal-use automobile sector, is advancing rapidly. Over the past decade the personal automobile fleet has been growing by over 20 percent a year.[2] Five factors suggest that future growth will continue at a high rate: (1) continued improvements in the Chinese economy; (2) the government's decision to make personal cars a pillar industry; (3) population growth and urbanization; (4) China's entry into the World Trade Organization (WTO); and (5) improvement and expansion of the transport infrastructure.

Currently, domestic production accounts for more than 95 percent of the total motor vehicle market share. Domestically produced trucks basically meet market demand in terms of variety and quantity. The previous high disparity between supply and demand in the passenger car sector has been largely relieved. China also has become the world's leading country in motorcycle manufacturing, boasting several internationally competitive motorcycle enterprises. Motorcycle output and product variety meet current domestic market needs.

The quality of vehicles produced in China is improving rapidly because of the tremendous growth in the production of vehicles by the joint venture companies, which provide their proprietary product design and modern factories. Chinese performance regulations are forcing the use of more advanced technologies. The competition introduced by China's entry into the WTO is expected to accelerate this technological and quality improvement.

CHINA'S AUTOMOTIVE INDUSTRY TODAY

By the end of 1999 China had 2,391 automotive enterprises: 118 OEMs, 546 motor vehicle remanufacturers, 136 motorcycle assemblers, 51 engine makers, and 1,540 motor vehicle/motorcycle parts and components companies (China State Economic and Trade Commission, 2001). China's automotive sector employed a total of 1.8 million people, of whom 169,000 were engineers and technicians. The automotive sector's total assets were

[2] Based on a presentation by Prof. Guo Konghui, Jilin University, Changchun, China, in Washington, D.C., May 2001.

TABLE 3-1 Motor Vehicle Production, China, 2000

	Production	
Vehicle Type	Volume (unit)	Percent Growth Rate
Vehicle total	2,069,069	13.07
Truck	764,005	1.03
Heavy	81,950	74.37
Medium	153,761	–16.82
Light	390,543	0.99
Mini	137,751	0.07
Bus	700,387	37.77
Large	7,953	3.06
Medium	35,938	22.13
Light	248,178	34.87
Mini	408,318	42.16
Car	604,677	6.95
2.5–4.0 liters	46,459	48.44
1.6–2.5 liters	367,554	2.97
1.0–1.6 liters	62,988	52.19
Less than 1 liter	127,676	–5.93

SOURCE: CAC-China Auto Consulting and CCPIT (2002).

RMB508.7 billion ($61.3 billion), and its total output value was RMB341.1 billion ($41.1 billion).

In 2000 total vehicle sales reached RMB391.1 billion ($47.1 billion), up by 80 percent from 1995, with profits of RMB17.4 billion ($2.1 billion), up by 107 percent from 1995. And in 2000 China produced 2.07 million motor vehicles (a 43 percent increase from 1995), 605,000 of which were passenger cars (an 86 percent increase over 1995), and 11.53 million motorcycles, or 44 percent of the world's total production (an increase of 45 percent over 1995). Overall, the automotive sector had a total export value of RMB20.7 billion ($2.5 billion) and an import value of RMB29.8 billion ($3.6 billion)—see Table 3-1 for a summary of 2000 production.

Finally, the annual production capacity for complete vehicles in China is about 2.6 million, including 1 million trucks, 700,000 buses, and 900,000 cars. The leading automotive manufacturing enterprises in China are listed in Appendix 3-A to this chapter.

THE CURRENT STATE OF VEHICLE TECHNOLOGY

Heavy-duty trucks. The Steyr series trucks, produced by imported technology, has reached the level of technology used by most international

OEMs of the 1980s. Today's Benz series trucks meet the 1990s technology standards. Vehicles produced by the Dongfeng Motor Corporation are meeting the international technology level of the late 1980s.[3]

Medium-size trucks. The medium-size trucks being developed and produced by domestic automobile manufacturing enterprises are relatively advanced in technology.

Light-duty trucks. Imported models of light-duty trucks include the Isuzu series and Iveco series, which have reached the international technology level of the late 1980s. The cabs of independently developed trucks, such as the Small Dongfeng and New Yuejin series, now meet the international level of the early 1990s. Some models developed in the 1960s and 1970s are still in production after modest improvements.

Minibuses and trucks. Most of these products were introduced from abroad in the mid-1980s, and there is a big gap in their technology compared with the advanced level reached internationally.

Large and medium-size buses. Imported series such as Kassbohrer, Volvo, Benz, and Neoplan are at the international technology level of the early 1990s.

Light-duty buses. The Iveco and Golden Cup series (Toyota Hiace), which have a large market share, have reached the international technology level of the mid-1980s. The newly imported Delica-Transit and Traffic series are at the level of the late 1980s.

Passenger cars. The Audi A6, Passat B5, and Bola models meet current international technology standards, as do the newly imported Buick Century and Honda Accord models. Fukang, New Jetta, and Audi are at the international level of the early 1990s. Alto and Cherokee are at the international level of the mid-1980s. The Santana series belongs to the products of an earlier period, but it has undergone major modifications.

Agricultural vehicles. Agricultural vehicles, which are low-speed and powered by diesel engines, carry both goods and passengers in rural areas.[4] The vehicles use very simple technology. In the late 1990s the annual production of such vehicles in China was about 3 million, most of which were three-wheelers. Prices vary from RMB4,100 ($490) to RMB28,700 ($3,460) per vehicle. Four-wheelers are capable of maximum speeds of about 50 kilometers per hour (kph) and payloads of up to about 1.5 tons, and three-wheelers have maximum speeds of up to 40 kph and payloads of up to 0.5 ton. Agricultural vehicles receive some tax advantages over conventional vehicles, but they are banned from most large cities because of their low speed and high pollution levels. Moreover, three-wheelers

[3] This section is based on CAC (2001).

[4] This section is based on Asian Strategic Investments Corporation (2000).

may be prone to overturning. At present China has about 15 million agricultural vehicles, produced by over 200 manufacturers. Virtually all companies are indigenous, using indigenous technology.

STRATEGIES PURSUED BY OTHER COUNTRIES TO DEVELOP THEIR AUTOMOTIVE INDUSTRIES

The U.S. and European automotive industries developed over many years, well before the global marketplace became a reality. In Europe many companies received direct government support from time to time, and most countries protected domestic producers at critical times. Geographically isolated, the U.S. industry developed with very little foreign competition until it was mature and very strong financially.

In the meantime, several research organizations were being established in Europe and the United States—for example, Ricardo (Shoreham by the Sea, Sussex, England), AVL (Graz, Austria), Southwest Research Institute (San Antonio, Texas, United States)—and today they provide both the smaller and the financially weaker companies, as well as the large, mature OEMs, with the help they need to develop technology. The annual budget of each of these research organizations is about $300 million (RMB2.5 billion). Another firm, entirely dedicated to automotive technology, is FEV of Aachen, Germany, with a laboratory in Detroit. Its sales in 2000 were $1.6 million (RMB13.2 million). Many universities have developed strong programs that have produced high-quality engineers and scientists in large numbers each year, effectively feeding the research programs both in industry and at independent laboratories.

Since World War II, Brazil, Korea, and Japan also have established large, internationally competitive automotive industries. Brazil has the twelfth largest automotive industry in the world. It produces 1.7 million vehicles a year, largely for the domestic market. Fiat, Volkswagen (VW), Ford, General Motors (GM), and Toyota each have freestanding, wholly owned subsidiaries in Brazil, which account for the production of the vast majority of vehicles. Each of these subsidiaries maintain some research and development (R&D) capability; in fact, the capability of VW's subsidiary is considered second only to that of the facility in Europe. Little intellectual property is transferred, but these enterprises provide jobs for Brazilians and contribute significantly to the country's gross domestic product (GDP). Although few strong university or independent automotive research centers exist in Brazil, some vehicles have been designed and developed largely by Brazilian researchers and engineers, including earlier versions of the Volkswagens now being manufactured and sold in China.

Japan and Korea, by contrast, have developed their own indigenous automotive industries—Japan beginning in the 1950s and Korea in the

1970s. In the early days in both countries, vehicles were built by large industrial companies with far-reaching industrial expertise and extensive resources. The Japanese and Korean governments provided R&D support, but the principal benefit they offered was protection from foreign suppliers. And to this day, Korean and Japanese companies account for over 90 percent of domestic vehicle sales in their respective countries. In both countries the early companies were export-oriented and able to establish a significant international presence within about 10 years of launching domestic models. Furthermore, both of these countries provided significant protection against foreign imports while their industries were developing, an option that is not available to China after its entry into the WTO. The Japanese industry also benefited greatly from having small, high-quality cars available when the market for small cars suddenly expanded. By contrast, there is no obvious technological capability in China from which Chinese industry might gain a major advantage that does not exist in foreign OEMs.

In Japan, one important exception to the industrial development just described is the Honda Motor Company. In the early years of the nation's automotive industry (1965), the Japanese Ministry of International Trade and Industry (MITI) decided that Japan should support only two or three automotive companies, and it developed its industrial policy accordingly (Johnson, 1982). Not one of the selected companies, Honda focused first on building its motorcycle products. Then, building on its expertise with motorcycle engines and technology, the company expanded in the 1960s to small cars. The company's first exports, in the late 1960s, were of poor quality and did not sell well. The next car it exported to the United States was the Honda Civic, with the innovative low-emitting, stratified charge CVCC (compound vortext-controlled combustion) engine. The timing was fortuitous because the United States was just implementing its first set of stringent emissions standards. Although other companies claimed that the standards were unreasonable and even unattainable, Honda disagreed and introduced its car to great acclaim. For several years, until the next series of more stringent nitrogen oxide (NO_x) standards were introduced, its CVCC engine was one of the few engines to meet U.S. standards without needing a catalytic converter. (About 15 percent of production in 1975 consisted of cars without converters, which included some produced by manufacturers other than Honda.)

The Japanese industry also benefited greatly from an unusual market circumstance in the 1970s and early 1980s. With the introduction of the oil embargo by members of the Organization of Petroleum Exporting Countries (OPEC), the demand for small cars with high fuel economy increased in the U.S. market. The Japanese imports were well suited to fulfill this demand, and U.S. domestic manufacturers offered essentially no competition for several years. During this time the Japanese companies captured a significant market share, one that they have continued to build on over the years. Although several international companies have announced intentions to produce vehicles in China for export, because China does not now make or export a unique vehicle, an indigenous Chinese industry is unlikely to enjoy similar export expansion.

THE FUTURE DEVELOPMENT OF CHINA'S AUTOMOTIVE INDUSTRY

Two important features of the 2001–2005 plan for the automotive industry are:

1. Two or three large automobile groups will be retained, with sales, distribution, and after-sales service systems commensurate with international standards. Their output will supply more than 70 percent of the domestic vehicle market and will include some exports.
2. The government will nurture the formation of 5–10 large supplier groups, which will compete in the international market. The largest three companies should enjoy a 70 percent share of the domestic market.

These guidelines represent a significant restructuring of the industry, from one that is heavily dependent on technology from outside sources—primarily from the world's leading OEMs that are partners in joint ventures—to one that is closer to being self-standing and capable of developing and introducing the latest technologies into Chinese products. In contrast with the once centralized control of the automotive industry, the more independent motorcycle and farm equipment industries have developed a fully stand-alone, indigenous capability that is able to compete successfully in the world marketplace.

In considering the implications of different industry structures, the committee identified various organizational forms and components that can be expected to be part of the total automotive industrial complex. Although it is not possible to predict accurately the size or complexity of any one of these elements, the overall structure of the industry will likely mirror the structure of the world industry. In a real sense, competition from successful overseas OEMs will encourage, perhaps even force, the Chinese industry to adopt many of their practices if it is to compete effectively in world markets. The elements of this future structure can be expected to include some or all of the following forms:

1. stand-alone indigenous OEMs
2. Chinese enterprises in partnership with a single distinct joint venture

3. Chinese enterprises, each in partnership with several joint ventures, including some overlap of joint venture members among the various Chinese enterprises, as is common today

4. motorcycle or farm equipment companies capable of developing a small car

5. wholly owned subsidiaries of foreign companies with the capability to manufacture new vehicles

6. small or modest-size domestic entrepreneurial companies able to provide engineering support to all of the various enterprises in China.

Although the first of the alternatives—stand-alone indigenous enterprises—has been assigned a definite role in the five-year plan, it is unlikely that current stand-alone companies will be able to achieve the level of capability needed without drastic restructuring. State-owned enterprises exist throughout China, but the committee was presented with no information that suggested that these have the organizational structure, the personnel, or the level of technological sophistication necessary to create world-class enterprises in the short term. Furthermore, state-owned enterprises are required to adhere to practices that increase their costs and make them less competitive, such as providing housing, schools, and facilities for their employees. Dramatic changes will have to be implemented in these enterprises if they are to provide a long-term basis for a truly competitive indigenous industry.

Several small entrepreneurial and locally owned enterprises appear to be emerging in the Chinese auto market. Although their cost structure seems to be competitive with the larger companies, the long-term viability of these enterprises remains unclear.

Anyone considering alternatives 2 and 3, both of which involve some form of a joint venture, must recognize that the technology contained in the world-class cars and trucks currently manufactured in China has been provided largely by joint ventures *without the transfer of the intellectual property* that would allow Chinese members of the joint ventures to develop their own capabilities. Intellectual property includes not only patents, but also a range of proprietary information such as trade secrets, special manufacturing techniques, design processes, and systems engineering. This type of knowledge is essential to the development of a competitive indigenous industry in China. In addition, as noted earlier, several of the vehicles produced by the joint ventures do not possess the latest technologies. It seems very unlikely that a competitive stand-alone capability can be achieved in the near term through the existing joint ventures without a drastic restructuring of these enterprises to ensure that the Chinese partner will gain the experience and know-how to design and develop world-class vehicles. This observation has been reported and reiterated in conversations with representatives of the major Chinese automobile companies and observed in the assembly plants.

The second form in the list—Chinese enterprises, each with a single joint venture—would have the benefit of fostering the kind of close working relationship between the joint venture partners that could allow the transfer of knowledge to and the building of a technical capability in the Chinese partner. With a single partner, intellectual property could be more easily protected, and the partnership could provide the structure that would allow the Chinese to achieve the long-term objective of developing an indigenous industry capable of competing in the world marketplace. Of course, if the joint venture partner is to be comfortable with creating that capability in its Chinese partner, it will want assurance that its Chinese partner will not become a competitor. The arrangement between the Chinese partner and the joint venture partner would, then, probably have to be a long-term one. But such a relationship is likely to be attractive to potential joint venture partners only if they foresee large markets for the products being developed by the enterprise or significant investment in the enterprise by the Chinese partners.

The third form—Chinese enterprises, each with several joint venture partners, some of which overlap with other enterprises—is the most common one at present, but it introduces a further complication. As noted earlier, the Chinese members of these enterprises have not benefited from the transfer of intellectual property from the foreign OEMs. Past business arrangements apparently did not require such transfers, and it is unlikely that joint venture partners would be willing to agree to such a structure in the future. Because each of the overseas members of the joint venture would want assurance that any proprietary information transferred to the joint venture would be kept confidential from other overseas members, who are likely to be competitors in other markets, the Chinese partner is likely to find it very difficult to receive and use this information to develop its own intellectual capabilities. Furthermore, this arrangement tends to reduce the incentive for the overseas partners to transfer the most sensitive proprietary information, which is precisely what the Chinese partners need to move toward a fully competitive independent industry that can design and develop a world-class vehicle.

The fourth form—the entry of either the motorcycle or the farm equipment manufacturers into the small car market—represents an interesting alternative for China. Many of these companies are freestanding, successful developers of products for their markets. Although there has been some indication that the motorcycle industry is interested in developing a small car, it appears that this vehicle is smaller than that contemplated for the "China car" described in the five-year plan. Because of limited information, the committee was able to conclude only that enterprises in these two industries represent a resource that could form an important part of the new automotive industry.

With China's entry into the WTO, the fifth form—wholly owned foreign subsidiaries in China—is likely to appear for the first time in modern history. Even some companies that are presently joint venture partners of Chinese firms may find that this is a more advantageous way to participate in the growing Chinese market. Already some major overseas component manufacturers—for example, Robert Bosch, Corning, and Engelhard—have wholly owned subsidiaries in China. Similarly, some overseas OEMs may choose to withdraw from joint ventures with a multitude of partners as a way of protecting their intellectual property. Such OEMs are more likely to participate in the market through wholly owned subsidiaries that will have access to the technologies of their overseas owners and can offer continuing competition to Chinese companies.

The sixth form of the industrial complex is the small or modest-size indigenous entrepreneurial company able to provide engineering support to all of the various Chinese enterprises. These independent engineering companies can provide important consulting and development capabilities. Large OEMs are increasingly turning to such resources in the West as they attempt to reduce their in-house costs. There is no reason why Western research companies would not be interested in supporting Chinese OEMs as well. As mentioned earlier, several of these companies have significant capabilities.

The principal difficulties that must be confronted by the Chinese companies attempting to develop the capabilities to design and develop an indigenous automobile can be summarized as follows:

- The Chinese industry must compete with strong, mature industries in Japan, Korea, Europe, and the United States.
- As a member of the WTO, China will find it difficult to protect its domestic industry as it enters the period when high growth in the domestic car market is about to accelerate.
- State-owned enterprises are not currently competitive with the major foreign OEMs.
- The enterprise structure that China has put in place through joint ventures with foreign OEMs, which brings several international companies together under one roof with domestic partners, will likely continue to discourage the transfer of knowledge to the Chinese because of fear that the knowledge will fall into the hands of competitors.

The critical issue the Chinese industry must face is how to develop a relationship with world-class OEMs that will encourage the transfer of the knowledge needed to build a long-term, independent, indigenous industry. As described earlier, companies could pursue several possible arrangements. Although all of these present possibilities for enhancing the viability of the Chinese automotive industry, it must be recognized that China's entry into the WTO is a strong countervailing force. With duties on imported vehicles expected to fall from 129 percent to 14 percent, some of the world's OEMs may view the opportunities of importing vehicles into China as more attractive than building the capabilities to design and develop vehicles in China. Inasmuch as the latter alternative will do little toward accomplishing the stated objective of building a freestanding, indigenous industry, the government will have to seriously consider its various alternatives.

The dilemma that China faces can be summed up in one question: Will China be able to revolutionize the structure of its automotive industry and encourage that industry to achieve a level of competitiveness at a time when much of the near-term market for small cars is affected by imports, thereby denying the local industry an important fraction of the revenue needed to develop the independent capability it is seeking? The movement toward rapid motorization does offer an incentive for restructuring the automotive industry, but the problems raised by the implications of WTO membership must be addressed. This overall challenge is embodied in a stated goal of the China's five-year plan for the automotive industry: to develop an economy car with a 1.3-liter engine that could be sold in China for about RMB80,000 ($9,600). Today in the United States at least two economy cars with a manufacturer's suggested retail price of less than $9,000 are currently for sale.[5] Both have 1.5-liter engines and 10-year/100,000-mile warranties and can be readily purchased for much less than the suggested retail price. With China's import tariffs for cars scheduled to be capped at 25 percent by 2005, such vehicles may offer serious competition in the economy vehicle segment of the Chinese market.

BUILDING AN R&D CAPABILITY

The creation of a strong domestic R&D capability is an essential element in the development of a successful indigenous automotive industry. This capability will enable both the transfer of intellectual property from international companies and the creation of intellectual property that can be the basis for a strong, export-oriented indigenous automotive industry.

Because many activities fall within the category of research and development, different companies define their R&D activities in different

[5] They are the Kia Rio and Hyundai Accent, both manufactured in South Korea.

ways, making it difficult to develop a clear picture of their efforts. For example, in the automotive industry the engineering involved in creating the product, the development of advanced technology, and the research on new concepts and procedures are commonly included under the rubric of research and development. If China hopes to develop an indigenous industry that can independently develop and sell vehicles that contain leading technologies in world markets, it must have these capabilities.

It appears from visits to various companies in China, presentations by experts, and visits to laboratories that such capabilities do not currently exist in China. China does possess several world-class manufacturing facilities and some scattered but excellent R&D capacity, both in and out of the industry, but its broad engineering and development capabilities relevant to the automotive sector do not appear to be comparable to those of the world-class OEMs. Currently, some 3,000 engineers are employed in research and development within the Chinese state-owned companies and the joint ventures involving foreign OEMs, but these engineers are not involved in the design and development of current vehicles. The reasons for these engineers' lack of involvement are not totally clear, but they appear to include some of the following.

Before the period of reform and the opening of the economy, R&D activities were limited through central planning. In the absence of competition, there was no incentive to either develop new vehicles or improve existing designs, and there was little opportunity for Chinese engineers to gain practical R&D experience. Today, in the post reform period, world-class cars are being produced in China, but the designs and the technology contained in them are imported by the joint ventures, and Chinese engineers are still given little opportunity to contribute to their designs. In its efforts to establish itself as a viable indigenous undertaking, the Chinese automotive industry must find a way to utilize this pool of talent successfully. Any restructuring within the industry must recognize this opportunity at the outset and make provision for it.

A reliable estimate of the resources needed to develop the required R&D capabilities would be helpful, but there is no generally agreed-on analytical procedure for accomplishing such a task. However, one measure of the annual investment needed may be the current annual investments in research and development of the world's competitive automotive companies. For example, according to documents filed by General Motors Corporation with the U.S. Securities and Exchange Commission, "in 2000, GM spent $6.6 billion [RMB55 billion] for research, manufacturing engineering, product engineering and development activities related primarily to the development of new products or services, or improvement of existing products or services, including activities related to emission controls, improved fuel economy, and the safety of persons using

GM products."[6] In 2000, GM spent 4.1 percent of its revenue of $161 billion (RMB1,300 billion) on research and development. The comparable figure for Ford was 4.8 percent. Product engineering and development consume the vast majority of R&D funds. (One of the Chinese enterprises informed the committee that it is currently investing 2 percent of sales in research and development, and that it is seeking to raise that figure to 2.5 percent. A second enterprise claimed to be aiming for an R&D expenditure of 3 percent of sales.) To compete with world-class manufacturers, China's automotive industry must be prepared to invest significant funds on a recurring basis. A critical mass of research and development may be required to effectively restructure the automotive industry. And during the time in which capability is being created, the funding will need to be significantly larger.

Two other critical issues also must be carefully considered. The first is the content of any planned R&D program. A wide spectrum of technologies is described in Chapter 4, some of short-term importance and some with long-term possibilities. Careful balance of the technologies to be explored in the R&D portfolio is essential. The industry cannot achieve the desired independence by concentrating exclusively on long-term or exclusively on short-term technologies.

Thanks to the sizeable investments made by members of the automotive industry worldwide, technology is continuing to evolve and new products are appearing almost daily. To remain competitive with these developments, the Chinese industry must possess the capability to contribute to current technological developments and to utilize them. Acquiring such a capability will require significant investments in both facilities and personnel. At the same time, the long-term technological opportunities cannot be ignored. China has undertaken steps to support some of the long-term technological research through its newly announced RMB880 million (over $100 million), five-year program for the support of research on fuel cells, hybrid vehicles, and electric vehicles. Although this investment is large by Chinese standards, it is only a fraction of what is being spent by overseas manufacturers to find solutions to the same technical problems. The challenge is to maintain a proper balance among the various technological opportunities.

A second critical issue, but one hard to quantify, is the availability in China of engineers trained in the various aspects of automotive engineering. Many new engineers with capabilities in the relevant fields are needed

[6] Form 10-K, as filed with the U.S. Securities and Exchange Commission, for the fiscal year ending December 31, 2000.

to join those already employed in industry. In addition, Chinese universities will probably have to increase their output of automotive engineers—only 178 masters degrees and 36 doctorates were awarded in automotive engineering in 2000—and some of the engineers receiving training abroad in this field should be encouraged to return.[7] But skills in product design and development are not acquired solely in school; industrial experience is essential. Engineers should be sent to as many international conferences, meetings, and short courses as possible to stay up to date on technological progress. The joint ventures should be encouraged to provide these opportunities, and state-owned enterprises and private companies with laboratories should participate in the training of students.

Overall, the Chinese government and the country's automotive industry will each have important roles in restructuring the industry to achieve its objectives.

CONCLUSION

The Chinese automotive industry is faced with a variety of structural problems. The state-owned industries must be made more efficient and all responsibilities not directly essential to automobile production should be removed and placed with other agencies. The joint ventures, while generally profitable, have been structured in ways that do not result in the transfer of technology to the enterprises. Re-creating a relationship with one or more joint venture partners in order to encourage the transfer of knowledge will be a challenge that may require restructuring some agreements among the joint venture partners. Independent entrepreneurial companies that can provide assistance to the industry are beginning to emerge as a source of technical expertise. The engineering consulting firms that have successfully supported the U.S. and European Union members of the industry represent a resource that can be utilized by the Chinese enterprises as well. Achieving the necessary restructuring while at the same time facing the likely competition from foreign imports (and possibly some freestanding subsidiaries of overseas OEMs) presents an enormous challenge to both the industry and the government. One thing is certain: creating a successful indigenous industry will require a long-term commitment of funding, substantial facilities, and the availability of highly trained engineers and scientists. A careful assessment of the resources likely to be available over time should be carried out before a particular path is chosen for the members of the industry.

---

[7] Information provided by Prof. Guo Konghui.

APPENDIX 3-A: THE MAJOR AUTOMOTIVE ENTERPRISES IN CHINA

FAW Group Corporation

FAW Group, the first large-scale motor vehicle production base in China, is now one of China's top ten industrial enterprises. Its headquarters is in Changchun, and its production capacity is 700,000 vehicles a year. In 2000 the total motor vehicle output of the FAW Group was 423,000, the highest in China's automotive industry. The FAW Group also produces mini- and light-duty buses and trucks, medium-size and heavy trucks, cars, and other series. Its Jiefang series buses and trucks constitute a large share of the market, and its Red Flag limousines are quite well known. The FAW Group also has entered into a joint venture with Volkswagen to produce Jetta and Audi sedans. Over the next five years the FAW Group is seeking to achieve a production capacity of 1 million vehicles, with a total sales volume of RMB820 billion ($10 billion).

On June 14, 2002, the FAW Group and Tianjin Automotive Industry Group Corporation (TAIC) jointly announced a merger agreement in which FAW will own 50.98 percent of Xiali Auto (CBU-AutoEnews, 2002a). TAIC also will transfer to FAW its 75 percent equity shares in the Tianjin Huali Automobile Company Ltd., a minivehicle manufacturer. With the acquisition of Xiali and Huali, FAW gains the only segment it missed—low-end, subcompact economy cars. As China's leading minicar and minivehicle manufacturers, Xiali and Huali bring to FAW not only two domestic brands, but also their national sales and distribution network. Moreover, the acquisition of Xiali gives FAW all the tangible and intangible assets needed to enter the economy car segment with the least capital, time, and risks. Over the past few years the FAW Group has expressed its intention to produce cars with an engine displacement of about 1.3 liters priced at about RMB80,000 ($9,600).

In August 2002 Toyota and FAW announced that they would join forces to produce luxury sedans, sport-utility vehicles, and minivehicles for the Chinese market—400,000 vehicles a year by 2010. Such a move will place the two automakers in direct contention with the Dongfeng and Shanghai groups and their international partners (Zaun and Leggett, 2002).

Dongfeng Motor Corporation (DMC)

Originally named the Second Auto Works, DMC has its headquarters in Shiyan City, Hubei Province. At present, DMC has three major production bases—Shiyan, Xiangfan, and Wuhan—which form the Hubei automotive industry corridor. In 1998 DMC produced 190,000 vehicles, mainly heavy-duty trucks, medium-size trucks, and light-duty trucks. Under a joint arrangement with the Citroën Corporation of France, it also will produce Fukang sedans. DMC is playing the leading role in achieving Hubei Province's goal of "building up a one million vehicle production base."

Beyond its role in Hubei Province, DMC has joined with Honda and Denway Motors in a venture to produce Honda cars in Guangzhou for export (*Wall Street Journal*, 2002). It also is negotiating a joint venture with Renault-Nissan (CBU-AutoEnews, 2002a).

Shanghai Automotive Industry Corporation (SAIC)

Shanghai began to manufacture cars in the 1960s, but on a very modest scale. In the 1980s it entered into a joint venture with Volkswagen of Germany to produce Santana sedans. By 2000 its production capacity had reached 400,000 vehicles and accounted for 45 percent of China's car market, with its profit exceeding the sum of that of all other automakers combined. As a result of a joint venture with General Motors, Buick Century sedans began coming off the assembly line at the end of 1998. The Shanghai Group also plans to develop its production of heavy-duty trucks, large buses, and light-duty vehicles. As of 2002 the Shanghai Group had established 44 joint ventures with global automotive companies.

China National Heavy-duty Truck Group (NHDTG)

NHDTG, the largest heavy-duty truck production group, mainly produces Steyr 91 series complete trucks, bus chassis, and engines for Germany's Mann Corporation, as well as transmissions, axles, and other products. In 1998 the group's total output was almost 9,000 trucks, and the company has set a production goal of 25,000 vehicles. In 2001 the group signed an agreement with Sweden's Volvo Corporation to commence production of heavy-duty vehicles.

Joint Venture Partners

Over the past few years, foreign companies have increased their participation in China's automobile industry. The current status of each joint venture partner is summarized in this section.[8]

---

[8] This section is based on a three-part series of articles entitled "World Automobile Giants' China Strategies in 21st Century" that appeared in January 2001 in the online version of China's *People's Daily* (2001).

Volkswagen

Shanghai-Volkswagen was established in 1985 and FAW-Volkswagen in 1991. Since then, the two ventures have sold more than 300,000 vehicles a year, maintaining their market share of over 50 percent. Volkswagen is trying to gradually stagger the products of its two ventures. The First Automobile Works (FAW, Changchun) put out the Audi A6 (C class) in 1999 and the Bora (A class) in 2001; it expects to produce minicars (A class) in 2004. Similarly, Shanghai produced the Passat (B class) in 2000, and family cars (A Class) were launched in 2002. The models retained their competitiveness with comparable products being offered elsewhere in the world.

All these models almost keep pace with the international market, and, when completed, the Volkswagen joint ventures will manufacture the top five models, based on overall production, in China. In addition, Volkswagen is studying the possibility of developing a new model to be sold in both China and overseas markets, and it may choose China as the base for export production. In fact, Volkswagen intends to introduce its most advanced manufacturing techniques and product technologies into China and bring its two Chinese ventures, as well as accessory systems, into its global orbit of purchasing, product research and development, and marketing.

Volkswagen remains the strongest producer of cars in China with total production of 412,127 in 2001 (*Wall Street Journal*, 2002). Shanghai Automotive Industry Corporation signed an agreement on April 12, 2002, with Volkswagen AG to extend their joint venture, Shanghai–VW Automobile Company, for another 20 years to 2030, according to a recent report (CBU-AutoEnews, 2002b). The total registered capital of the joint venture will increase to RMB6.3 billion ($760 million) from the current RMB4.6 billion ($550 million).

General Motors Corporation

Since 1989, General Motors, the world's largest automobile manufacturer, has invested about $2 billion in China to set up three vehicle joint ventures—Shanghai GM, Shenyang Gold Cup GM, and Liuzhou Wuling Motor Company—and one solely funded accessory sales center. General Motors has successfully brought a series of its products into China, including the Buick sedan, the Buick GL8 for business, the Sail family car, Chevrolets, and pickup vehicles. Because it plans to turn Shanghai GM into its production base in Asia, General Motors has provided the Pan Asia Technical Automotive Center in Shanghai with major support. GM's overall production of the Buick Century, Buick GL8, and the Sail climbed to 59,729 in 2001(*Wall Street Journal*, 2002).

Ford Motor Company

Ford, which was the first American automobile manufacturer to enter the Chinese market, in June 1978, currently has in China more than 10 sales agencies, over 40 service facilities, and 2 global accessory sales agencies, as well as a technology training center. Its Transit vehicle, codeveloped with Jiangling of China, is now in production. Currently, Ford-brand vehicles control 0.5 percent of the Chinese auto market share, according to the company (Xinhua News Agency, June 18, 2002).

Chang'an Ford is Ford's first passenger car joint venture with Chang'an Motor Corporation based in Chongqing in southwest China. Ford will introduce a family-size sedan, which is based on its Fiesta platform developed in Europe, into the 50/50 joint venture. The $98 million (RMB810 million) joint venture was approved by the Chinese government during the first half of 2001. In preparation for selling both vehicles that it will produce in China and imported Ford-brand autos, the joint venture has selected 25 franchise dealers in China. Ford said earlier that it also plans to set up an auto financing branch to serve local consumers before the first product launch of the joint venture.

DaimlerChrysler

Sales of DaimlerChrysler's Chinese product, the Beijing Jeep, declined throughout the 1990s, revealing the need for urgent technological improvements. On September 27, 2000, together with Beijing Automotive Industry Group, DaimlerChrysler declared that it would invest an additional $226 million (RMB1.9 billion) to strengthen and expand production in China. DaimlerChrysler also has joined hands with Yaxing-Benz (Yangzhou) in bus production, obtained approval to manufacture trucks in Baotou, Inner Mongolia, and signed a technology transfer agreement with Ankai Auto (Anhui) to produce a luxury car. It continues to produce the Jeep Cherokee in Beijing with total production in 2001 of only 4,258 (*Wall Street Journal*, 2002).

PSA Peugeot-Citroën Group

PSA began to enter the Chinese market in the late 1980s. For various reasons, its sedan project in Guangzhou produced only 100,000 cars in 10 years. Then PSA withdrew its funds, resulting in the total collapse of the project, which was taken over by Honda (see section on Honda). PSA's Shenlong-Citroën joint venture with Dongfeng Motor Corporation also experienced many setbacks, including the 10 full years required to commence operations. In September 2001 Shenlong added RMB3.41 billion

($411 million) to its registered capital via a RMB2.34 billion ($282 million) debt-to-equity transfer, bringing total capital to RMB6 billion ($720 million). Meanwhile, the French shareholders of Citroën added an amount that kept its share at 30 percent, and Dongfeng's share declined to 31 percent. In 2001 Citroën's Fukang (ZX) and Picasso models led the company to being the third largest car producer with 53,680 units (*Wall Street Journal*, 2002).

Renault

Renault and the Sanjiang Group in Hubei Province joined forces in 1993 to assemble a light bus, Trafic. But production was suspended because of poor sales after the venture sold only 4,906 Trafic vans in seven years. Renault also has talks under way with Beijing Automotive Industry Group about producing the "Scenic" sedan and with Dongfeng Motor Corporation on trucks. As noted earlier, DFC is reportedly in the final stages of negotiating a joint venture with Renault-Nissan (CBU-AutoEnews, 2002a).

Toyota

Japan's biggest auto corporation, Toyota, had plans to produce automobiles in Tianjin in 2002. Its Xiali, coproduced by the Daihatsu Motor Corporation and Toyota's partner, the Tianjin Automotive Industry Group Corporation, had dominated China's taxi market. However, in the three years from 1999 to 2001 TAIC's car market share in the country fell drastically, from 18.5 percent to below 10 percent. Xiali had been steadily losing its market share with taxi fleets because large and medium-size cities were choosing larger taxi models. The availability of other subcompact cars such as the Yueda-Kia Pride, the Nanya Eagle, and the SAIC-Qirui Chery also drew customers away from the older Xiali model. At the higher end, the successful promotional activities of the Buick Sail, even though launched six months later than TAIC's new Xiali 2000, overshadowed the latter's launch. In 2001 the Xiali 2000 sold only 10,000 units compared with the 28,000 units sold of the Sail.

As noted earlier, Toyota and FAW announced in August 2002 that they would jointly produce 300,000–400,000 luxury sedans, sport-utility vehicles, and minivehicles a year by 2010. (Zaun and Leggett, 2002).

Honda

Honda's well-known motorcycle engine technology has earned it an important position in China's motorbike market. Its other success is Guangzhou Honda, which will produce a new model each year, including a new minicar. Guangzhou Honda had stated its intention to raise its annual production of Accords from 30,000 to 50,000 units by 2002, but was already exceeding that level in 2001, producing 51,116 Accords. More recently, Honda also announced that it plans to build a factory in China where it will make cars exclusively for export to markets in Asia and Europe (*Wall Street Journal*, 2002). The new plant will be located in Guangzhou and will be operated in partnership with two Chinese auto makers—Guangzhou Auto Group Corporation, owned by Hong Kong-listed Denway Motors Limited, and Dongfeng Motor Corporation. In the same announcement, Honda said that it plans to increase annual capacity at its existing plant to 120,000 vehicles by March 2003.

The Honda Odyssey, a multipurpose vehicle (MPV) model made by Guangzhou Honda, rolled off the production line on April 10, 2002. The retail price of the Odyssey is RMB298,000, or $36,000 (CBU-AutoEnews, 2002b).

Nissan

Nissan joined Zhengzhou Light Vehicle Factory in 1994 in manufacturing pickup trucks, but output remains low. It also joined with Yulon Motor Company and Dongfeng Motor Corporation to produce the Fengshen Bluebird in Shenzhen. Finally, Nissan has entered into a partnership with Renault for sedan manufacture in China. Specific models and investment partners have yet to be announced.

Hyundai Corporation

Early in 2000 Hyundai Corporation signed a letter of intent to expand cooperation with the Jiangsu Yueda Group, and in September 2000 it formally signed an agreement on transferring stock rights and making additional investments to set up a joint venture with equal shares by both sides.

In 2001, 7,715 Hyundai-Kia Pride vehicles were produced in China (*Wall Street Journal*, 2002).

According to the state press, China has picked a car made jointly with Hyundai as its preferred vehicle to replace Beijing's vast and often dilapidated taxi fleet in preparation for the 2008 Summer Olympics (Xinhua News Agency, July 24, 2002). The vehicle, the midsize Sonata Saloon, will be produced by a 50/50 joint venture of Hyundai and Beijing Automotive Industry Corporation.

Daewoo

To qualify for sedan manufacture, Daewoo established two large ventures in China in the mid-1990s: FAW-Daewoo Automotive Engines and

Shandong-Daewoo Auto Parts and Components Company. Daewoo's bankruptcy in 2000 adversely affected the two projects, as well as a bus project in Guilin. The Shandong project, which still lacks central government approval to assemble vehicles, is at a standstill.

APPENDIX 3-B:CHINA'S FIVE-YEAR PLAN FOR THE AUTOMOTIVE INDUSTRY—GOALS AND STRATEGIES

China's five-year plan for the automotive industry identifies concrete goals and strategies for use in stimulating the rapid growth of the industry. The specific goals that affect the automotive industry are listed below:

- Total annual vehicle production of 16.2 million units is sought, of which passenger cars will be 1.1 million units and motorcycles 13.0 million units.
- Two to three large automobile groups will be formed, with a sales, distribution, and after-sales service system commensurate with international standards. Their output will supply more than 70 percent of the domestic car market and will include some exports.
- The government will nurture the formation of 5–10 large supplier groups, which will compete in the international market; the largest three companies should enjoy a 70 percent share of the domestic market.
- The country also will form three or four large motorcycle group corporations, which will be highly competitive in the international market.
- The number of diesel-powered trucks and buses will increase, with all medium-size vehicles powered by diesel engines. The production of diesel-powered cars and minivehicles also will be initiated. Output of diesel vehicles will increase from 29.7 percent to over 35 percent.
- Output of alternative fuel buses and taxis also should increase, to about 2 percent of total output.
- Cars equipped with carburetors and using CFC-12 (chlorofluorocarbon) as a refrigerant will be discontinued.
- Safety features will significantly improve, with antilock braking systems (ABS) applied to large and medium-size buses and heavy-duty trucks. More passenger cars will be equipped with ABS and air bags, and new cars and light and minibuses will satisfy side-impact requirements.
- New cars, light and minivehicles, large and medium-size buses, and heavy-duty and medium trucks should meet European Emission Standard II (Euro II) emissions standards. Mid- to high-level cars and luxury large and medium-size coaches should meet Euro III emissions

standards. New types of four-wheeled farm vehicles powered by multicylinder engines should meet Euro I standards. China is aiming to achieve international emissions standards by the year 2010.

- Average fuel consumption rate of new cars and light vehicles will be reduced during the plan period by 5–10 percent and that of heavy-duty and medium-size trucks by 10–15 percent.
- For passenger cars, emphasis will be on developing economy models that have engine displacement of 1.3 liters and that are smaller and priced around RMB80,000 ($9,600). Such economy models should meet national safety, fuel efficiency, and emissions standards and the demand of individual consumers.
- For taxi models, environmentally friendly cars should be developed and used.
- Balanced development is needed for passenger car diesel engines, single-fuel compressed natural gas (CNG) and liquefied petroleum gas (LPG) engines, and hybrid power systems that meet Euro II and Euro III emissions standards.
- For trucks, emphasis will be put on heavy-duty vehicles with engine displacement of 9 liters or more and horsepower of 300 and up for use on expressways. These include large horsepower tractor trailers, heavy-duty special-purpose vehicles, and chassis and diesel engines with 300 and higher horsepower that meet Euro II and III emissions standards.
- On the basis of current manufacturing conditions and facilities, light and minivehicles for use in the countryside will be developed.
- Emphasis will be placed on the development of environmentally friendly city buses.
- The focus on new motorcycle development will be on environmentally friendly motorcycles (such as fuel injection and electric models), engines that meet new emissions standards, and key parts and components. In the meantime, the industry should develop new, reliable, inexpensive models that are easy to repair and that suit rural road and loading conditions. The government will continue to support motorcycle exports.
- The proportion of electronic products in a motor vehicle and the use of high-performance, lightweight, energy-saving, and environmentally friendly materials will be expanded.
- Research and development related to electric and hybrid vehicles will be increased.
- Research and development related to the use of recycling materials for environmental protection will be expanded.
- Additional enterprises will be supported in their efforts to set up modern sales and distribution systems that combine sales, parts supply, service, and repair and information feedback using the Internet.

• Guidelines and principles for fair competition, opening to the outside world, and fostering independent development capability while seeking international cooperation will be developed.

• A unified national policy to expand the domestic automobile market will be created.

• National standards will be developed for car purchase and use by eliminating excessive fees; simplifying the procedure for car purchase, registration, and use; and increasing the availability of car purchase loans.

• For nationwide fuel economy and promotion of economical cars, a comprehensive fuel tax system will be implemented.

• The official used vehicle supply system will be reformed.

• Economic and technical cooperation with international partners will be encouraged.

• Fuel quality will be improved.

• Local governments will be responsible for roads, parking, and other infrastructure. They also will have authority to improve traffic management and road capacity.

REFERENCES

Asian Strategic Investments Corporation. 2000. China's Agricultural Vehicle Industry: A Market Survey. Beijing, June.

CBU-AutoEnews (China Business Update). 2002a. 8(June 27).

———. 2002b. 3(April 18).

China National Automotive Industry Consulting and Developing Corporation (CAC). 2001. China's Automotive Industry: Its Past, Present and Future. Online. Available at www.cacauto.com.report-en/report8.htm. Accessed October 10, 2002.

China National Automotive Industry Consulting and Developing Corporation (CAC-China Auto Consulting) and Economic Information Department, China Council for the Promotion of International Trade (CCPIT). 2002. Production and sales of motor vehicles in January–December 2000. Online. Available at www.cacauto.com/databank/prsa00121.htm. Accessed August 28, 2002.

China State Economic and Trade Commission. 2001. China's 10th Five-Year Plan for the Development of the Automotive Industry (2001–2005). Translation made available by Volkswagen Corp. June.

Johnson, C. 1982. MITI and the Japanese Miracle: The Growth of Industrial Policy, 1925–1975. Stanford: Stanford University Press.

People's Daily. 2001. World automobile giants' China strategies in 21st century. Online. Available at http://english.peopledaily.com.cn/200101/12/eng20010112_60306.html; http://english.peopledaily.com.cn/200101/15/eng20010115_60493.html; http://english.peopledaily.com.cn/200101/16/eng20010116_60600.html. Accessed May 13, 2002.

Wall Street Journal. 2002. Honda aims to export cars from China. July 11.

Zaun, T., and K. Leggett. 2002. Toyota plans to produce luxury vehicles in China. Wall Street Journal. August 30.

4

Present and Future Automotive Technologies

The tenth five-year plan envisions a future for China in which cars will be widely available to Chinese families and in which the Chinese automotive industry will grow into a "pillar industry" of China's economy. Technology will play an important role in facilitating these goals.

This chapter describes the automotive technology options that are available today and some that may become available in the longer term, and it comments on their applicability to the development of China's automotive industry and road transportation fleet. The choice of automotive technology is closely related to the kinds of materials and fuels selected, the economic impacts of the technology, and infrastructure requirements. The development of substantial transportation infrastructure, which requires major investments in land, influences in turn how land is developed. These issues are complex and variable, strongly depending on local values and conditions (see Chapter 6 for more discussion about the effects of motorization on land use).

Because the car is part of a much larger system, the overall costs and benefits of each technology choice must be assessed on a system basis. Such an assessment may reveal that the optimization of a single component may not be the best choice when the total system is considered. This life cycle analysis has been used in technology assessment studies worldwide and is appropriate for strategic decision making.

Although this chapter focuses on automotive technologies, many of the issues considered elsewhere in this report—automotive industry issues (Chapter 3), energy/fuels issues (Chapter 5), societal change (Chapter 6), environmental and health concerns (Chapter 7), and government policies (Chapter 8)—impinge on the choices of technology. China has already decided to require Chinese cars to meet European Emission Standard I (Euro I) now and Euro II standards by 2004–2005, based on state environmental protection regulations. The European Union[1] has been enacting increasingly stringent emissions standards and now requires Euro III standards for its member nations, with Euro IV standards planned for implementation in Europe in 2005. State plans call for Chinese cars to attain the current Euro emissions standards by 2010.

To meet the "wide availability" objective (that is, be affordable), the "China car" would have to cost about RMB60,000–80,000 ($7,200–9,600). Below this price range, the basic requirements for emissions control, performance, and safety cannot realistically be met.

To make the automotive industry into a pillar of the economy, China will have to build its domestic capabilities for supplying materials, manufacturing components, and providing the design, assembly, maintenance, sales, credit, and other services that in turn will contribute to an increase in China's gross domestic product (GDP). As discussed in Chapter 3, consolidation of the Chinese automotive industry and improvements in efficiency and cost competitiveness will take on new importance. Imported cars and components create unfavorable trade balances for China, whereas exports will expand opportunities for its automotive industry. Where technology choices bear on the future competitiveness of the industry, these factors must be considered carefully.

Likewise, the China car should be highly fuel-efficient, because China's domestic supply of petroleum fuel is limited at present and higher fuel demand can be met in the near future only through increased imports. This situation will affect China's balance of trade over the next decade and perhaps longer. The China car also should be highly "clean" in emissions, because the air quality in many Chinese large cities ranks among the worst in the world. The situation is expected to worsen if vehicle tailpipe emissions are not controlled rigorously.

The buyer of the China car will have a heavy influence on automobile design and sales. Successful auto companies select product attributes that maximize markets. Consumers typically want the most performance and convenience they can afford. Initially, most Chinese drivers will be located in urban areas or on rural and intercity roads that are not designed for high speeds. The performance requirements and fuel efficiency of cars are heavily influenced by the "driving cycles" used in designing and op-

[1] Through rulemaking proposed by the European Union Council of Environmental Ministers and approved by the European Parliament.

erating the vehicle. In the United States, where motorists engage in a lot of high-speed highway driving, many cars are designed with large engines to enable passing and hill-climbing capability at high speed. But these cars have poor fuel efficiency at low speeds and continue to burn fuel at a high rate when idling in urban traffic jams. Cars designed specifically for urban driving cycles could be much more fuel-efficient under these conditions. For example, new technologies based on greater electrification of the car are being developed to switch the engine off during vehicle stops. This feature has the double benefit of reducing emissions and wasted fuel. The envisioned China car also will have to be highly reliable, easily serviceable, and rugged to accommodate China's present road and maintenance infrastructure.

Over the century of development of automobiles in member countries of the Organisation for Economic Co-operation and Development (OECD), governments and automakers continually responded to problems and applied the lessons learned to the development path. For example, pollution problems led to increasingly stringent emissions regulations; safety problems led to improved roadway designs, crashworthy vehicle designs, and safety features such as seat belts and air bags; and operational problems led to driver training and vehicle inspection programs. China, facing an anticipated period of rapid investment and growth for automobiles and the associated infrastructure of a decade or two rather than a century, will benefit from those lessons as well.

The technology choices that must be made in China are constrained by the marketplace and by various government policies and regulations. With that in mind, this chapter will focus on technologies that seem to best fit the rapidly growing automotive industry in China over the next five years or so, and will examine them within the context of both their short- and long-term implications for the industry and consumers. The discussion of near-term technologies will be placed in the context of current expectations of government policies. The discussion of longer-term technologies will address those that could be important whether current government policies remain as they are or not.

FLEET ISSUES FOR VEHICLE TECHNOLOGIES

Emissions Control

In view of the projected growth of China's vehicle fleet, air quality issues in major urban areas are of great importance. As just noted, Chinese cars produced in 2004–2005 must meet Chinese emissions standards equivalent to the Euro II standards. China's Technical Policy on Prevention and Control of Motor Vehicle Pollution states that emissions levels

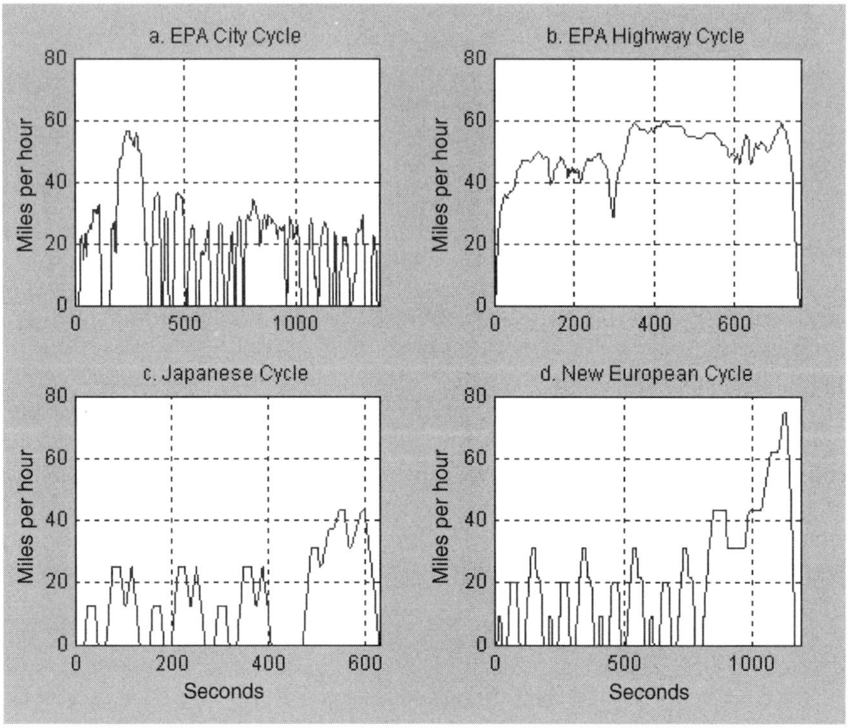

FIGURE 4-1 Driving cycles for measuring emissions. NOTE: 20 miles per hour (mph) = 32 kilometers per hour (kph); 40 mph = 64 kph; 60 mph = 97 kph; 80 mph = 129 kph.

should approach the international level of control around the year 2010 (Chinese State Environmental Protection Agency, 1999). Table 4-1 summarizes the principal elements of the existing U.S., Japanese, and European standards. Emissions control technologies to meet these standards are discussed later in this chapter.

Vehicle emissions are the products of the incomplete combustion of propulsion fuels. Propulsion system efficiencies vary with output energy requirements, and different propulsion technologies achieve optimum efficiency under different conditions. In urban driving, conventional engines continue to run, using energy and producing emissions, even if a vehicle is stopped or moving slowly in traffic. Because any comparison of vehicle options must take into account the typical local driving conditions, when emissions standards are set they are based on a specified "driving cycle." Figure 4-1 presents four examples of driving cycles specified by regulatory groups in the United States, Japan, and Europe.

TABLE 4-1 Examples of Emissions Standards in the United States, Japan, and Europe

	Date of Implementation	CO	HC	NO_x	PM
U.S. Standards					
SI car	1997	3.4 g/mi (2.1 g/km)	0.41 g/mi (0.26 g/km)	0.40 g/mi (0.25 g/km)	N.S.
SI and diesel cars	2007	2.1 g/mi (1.3 g/km)	0.09 g/mi (0.06 g/km)	0.07 g/mi (0.04 g/km)	1.01 g/mi (0.006 g/km)
Japanese Standards					
SI car	2000	0.67 g/km	0.08 g/km	0.08 g/km	N.S.
Diesel car	2002	0.63 g/km	0.12 g/km	0.30 g/km	0.056 g/km
European Standards					
Euro II SI car	1996	3.28 g/km	0.34 g/km	0.25 g/km	N.S.
Euro II diesel car	1996	1.06 g/km	0.19 g/km	0.73 g/km	0.10 g/km
Euro III SI car	2000	2.30 g/km	0.20 g/km	0.15 g/km	N.S.
Euro III diesel car	2000	0.64 g/km	0.06 g/km	0.50 g/km	0.05 g/km
Euro IV SI car	2005	1.00 g/km	0.10 g/km	0.08 g/km	N.S.
Euro IV diesel car	2005	0.50 g/km	0.05 g/km	0.25 g/km	0.025 g/km

NOTE: CO = carbon monoxide; HC = hydrocarbons; NO_x = nitrogen oxides; PM = particulate matter; SI = spark ignition; g/mi = grams per mile; g/km = grams per kilometer; N.S. = no specification.

SOURCE: Yasuhira Daisho, Waseda University.

In the United States, emissions are set on a per mile basis using a combined urban-highway cycle. Japanese and European driving cycles are more similar to the U.S. urban driving cycle than to the U.S. highway driving cycle. A comparison of the urban driving cycles used in Japan (Figure 4-1c), Europe (Figure 4-1d), and the United States (Figure 4-1a) to measure emissions indicates that in the Japanese and European cycles more time is spent with the vehicle stopped, engine idling, than in the U.S. cycle, which has an higher average speed and less idling time. Figure 4-1b depicts the U.S. highway cycle that was used in combination with the U.S. city cycle to determine vehicle fuel economy (45 percent share of the highway cycle and 55 percent of the city cycle).

When vehicles are operated only in urban environments where traffic constrains their rate of acceleration and maximum speed, users have modest expectations of vehicular performance in terms of acceleration and top speed. If users are able to drive vehicles on improved roads at high speeds with little competing vehicular traffic, their expectations for engine power and high-speed capability will rise. Because China has chosen the Euro emissions standards, it will have to measure emissions performance for the driving cycle that is established for the Euro system. In the future, however, China may choose to implement new standards, with a driving cycle and emissions standards suited to its specific environment.

Energy Use and Fuel Economy

The energy use of a vehicle fleet depends on the size, weight, type, and efficiency of vehicles in the fleet and on the driving conditions encountered in their use. Energy use includes not only the fuel consumed in operating the vehicle, but also the energy consumed in making the vehicle, producing and processing the fuel, and disposing of the vehicle at the end of its life.

Life Cycle Assessment

Anyone weighing the automotive technology options discussed later in this chapter would benefit from comparing alternative automotive technology systems over their full life cycle (see Figure 4-2, which depicts the steps in the life cycle of automotive technology from the production of the raw materials used to make the fuel and the vehicle through the vehicle's useful life to its final disposition). Such an assessment would allow one to track the key parameters involved through each of these life cycle stages and assess the overall results as part of the technology selection process. For the vehicle, the major parameters of interest are: cost, performance, local emissions of air pollutants, greenhouse gas (GHG) emissions, and

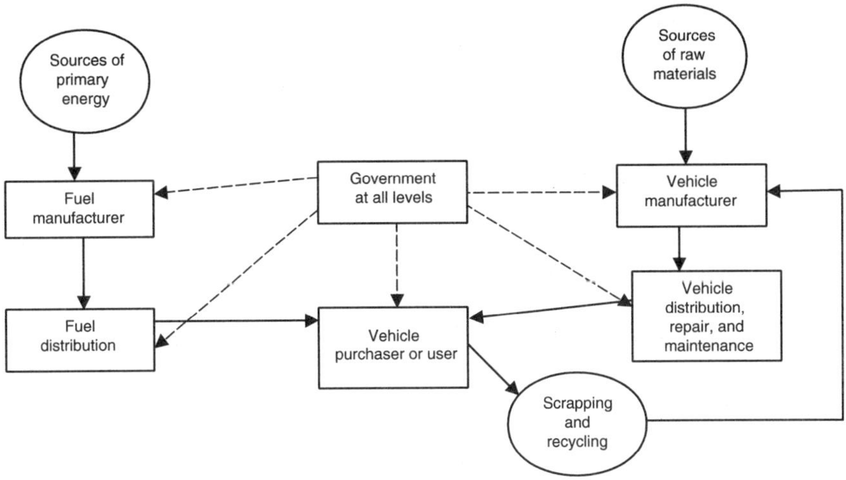

FIGURE 4-2 Steps in the life cycle of automotive technology. SOURCE: Weiss et al. (2000).

energy use. The vehicle also must have attributes that make it attractive to its purchasers and users over its lifetime, and it must meet established regulatory standards.

Various recent studies have compared the technologies for clean, fuel-efficient cars (Office of Technology Assessment, 1995; *Automotive Engineering*, 1996; Sierra Research, 1997; Höhlein et al., 1998; Singh et al., 1998; Ogden et al., 1999; NRC, 2000; Pembina Institute, 2000; Weiss et al., 2000; GMC, 2001). Not all of these studies treat the full life cycle of comparable fuel/vehicle systems. For example, the U.S. Partnership for a New Generation of Vehicles (PNGV) program does not consider the fuel cycle in its performance goals or comparisons of vehicles (NRC, 2000). And the General Motors study does not provide information on the production costs of new vehicles, nor does it state detailed design assumptions about the vehicles evaluated. The results of the study by Weiss et al. (2000) are summarized here to illustrate the importance of a life cycle review.

Beyond the organizations directly involved in producing fuels and vehicles are the vehicle purchasers and various levels of government. Those purchasing vehicles make their choices based largely on affordability, convenience, comfort, availability, and appearance. Local governments impose local zoning and safety codes; subnational governments issue planning, tax, and regional environmental regulations; and ultimately the national government is responsible for central investments in infrastructure, for national tax policies, and for the national trade, environmental, safety, and other requirements that are applied to fuels and vehicles.

In the life cycle assessment used here for illustrative purposes (Weiss et al., 2000), the starting point is a vehicle similar to a 1996 Toyota Camry. It is assumed that this vehicle, the baseline vehicle for the study, evolves forward to the year 2020 into a car of similar performance and capacity. The analysis is based on a U.S. Environmental Protection Agency (U.S. EPA) combined city-highway driving cycle. Various choices in technology are then considered, resulting in the following set of study cases:

- Baseline vehicle—the 2020 version of a car similar to a 1996 Toyota Camry, with a gasoline engine, 600 kilometer (km) refueling range, and some body lightweighting
- Advanced body vehicle—a similar car that is about 10 percent lighter than the baseline because of changes in materials and that costs about 10 percent more
- Advanced body, diesel (with both petroleum-based fuel and Fischer-Tropsch synthetic diesel fuel made from natural gas)
- Advanced body, hybrid—gasoline, diesel, or compressed natural gas (CNG)
- Advanced body, fuel cell hybrid—dependent on reforming gasoline, reforming methanol, or utilizing high-pressure hydrogen gas made from natural gas
- Advanced body, electric (requiring recharging every 400 km because of battery limitations).

The comparisons that follow are based on these vehicle designs. But because forecasting technology some 20 years in the future involves some uncertainty, the forecasts here related to evolutionary and advanced body combustion engine cars are subject to underestimation or overestimation by about 10 percent; the hybrids by about 20 percent, and the fuel cell and electric vehicles by about 30 percent. The uncertainties are greater for the rapidly evolving technologies because of the possibilities of technological breakthroughs or the identification of unanticipated barriers. The study by Weiss et al. (2000) assumes a vehicle lifetime of 15 years and an annual distance traveled of 20,000 km.

Life Cycle Energy Consumption

Figure 4-3 shows the life cycle energy consumption, on a megajoule (MJ) per kilometer basis, of the various technology combinations evaluated. The top portion of each bar represents the energy equivalent of the fuel used by the vehicle in propulsion, the middle portion the energy used

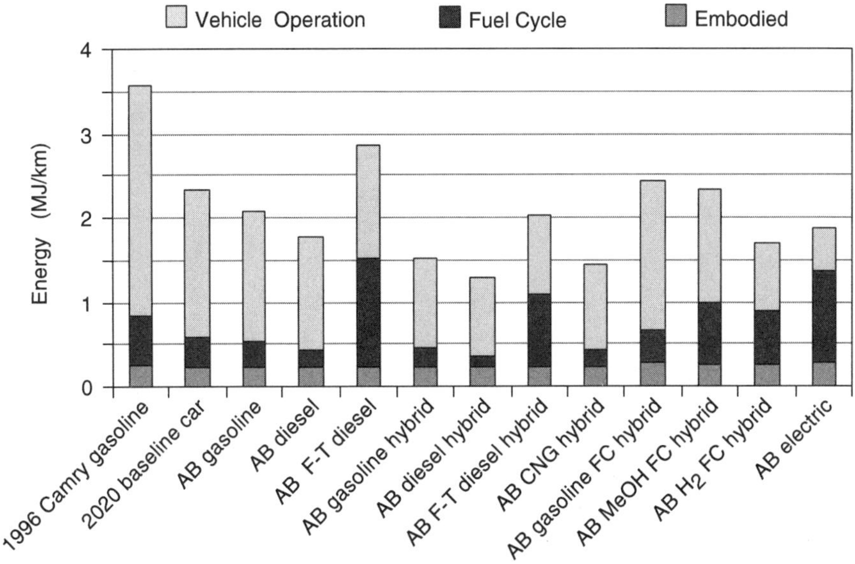

FIGURE 4-3 Comparisons of life cycle energy use. NOTE: MJ/km = megajoule per kilometer; AB = advanced body; F-T = Fischer-Tropsch; CNG = compressed natural gas; FC = fuel cell; MeOH = methanol; H_2 = hydrogen. SOURCE: Weiss et al. (2000).

in producing that amount of fuel, and the bottom portion the energy used in the actual manufacture of the vehicle (embodied energy). For the cases studied, the energy involved in vehicle manufacture is a relatively small part of the life cycle energy use. For much smaller, more efficient vehicles, the embodied energy becomes a more significant factor.

For petroleum fuels and natural gas, the energy required for fuel production is a relatively small part of total energy requirements (Figure 4-3). However, the energy associated with producing a synthetic fuel by a Fischer-Tropsch process adds substantially to life cycle energy use, which is related to GHG emissions as well. The energy production requirement for electric vehicles is based on the mix of energy sources and power generation efficiencies of the U.S. electricity supply. On that basis, the fuel cycle is the predominant energy requirement for the electric vehicle.

Life Cycle Costs and Emissions

Figure 4-4 presents similar comparative data, but it looks at total annual operating costs, new vehicle costs, and total GHG emissions. It is assumed that the vehicles would have to incorporate technology to meet

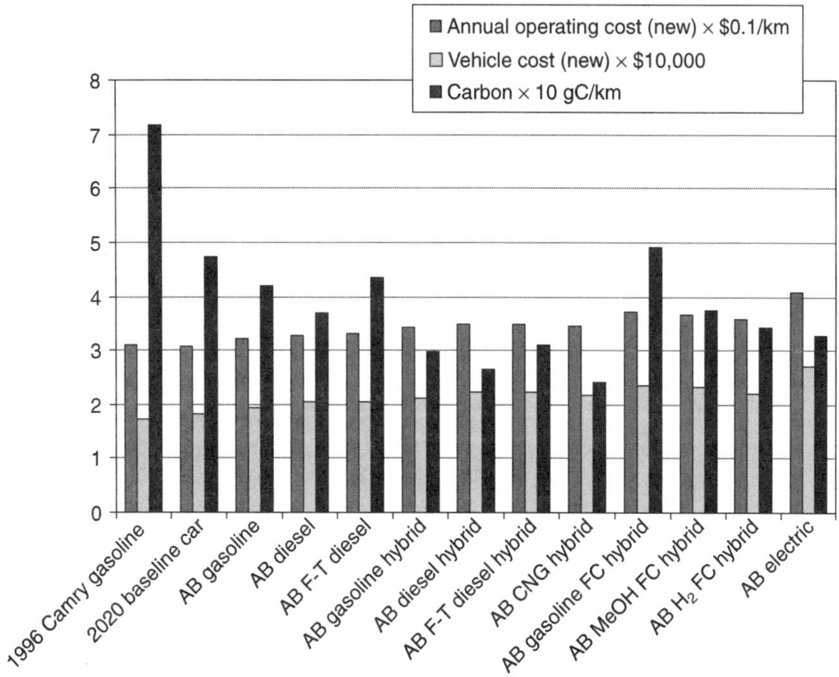

FIGURE 4-4 Life cycle comparisons of costs and carbon emissions. NOTE: Annual operating costs include amortized new vehicle costs and running costs as shown in Table 4-2. AB = advanced body; F-T = Fischer-Tropsch; CNG = compressed natural gas; FC = fuel cell; MeOH = methanol; H_2 = hydrogen; gC/km = grams carbon per kilometer. SOURCE: Weiss et al. (2000).

the 2020 U.S. emissions standards for local pollutants. The pattern of GHG emissions follows that of energy use for the petroleum fuel-based systems, but emissions are somewhat reduced for the CNG-fueled hybrid. GHG emissions for the hydrogen fuel cell and for the electric car are slightly higher than those for the gasoline hybrid, based on the U.S. electric sector average GHG emissions.

Table 4-2 presents estimates of the annual operating costs for new U.S. vehicles based on fuel cost averages and vehicle fuel consumption. A flat fuel tax of $0.0033 (RMB0.027) per megajoule of fuel ($0.40 per gallon of gasoline equivalent) is used across all the fuel sources (this assumption is made so that tax policy does not affect relative results; taxation is a policy tool that may be used to influence the economic choice between technologies). A constant maintenance or other charge of $0.036 (RMB0.30) per kilometer for the various technologies is assumed to avoid introducing an additional bias. Total fixed costs are based on known fixed annual

TABLE 4-2 Vehicle Costs per Kilometer for Selected New Vehicle Options, 2020 (1997 U.S. dollars per kilometer)

	Evolved Body Gasoline	AB Gasoline	AB Diesel	AB Gasoline Hybrid	AB Diesel Hybrid	AB, CNG Hybrid	AB Gasoline FC Hybrid	AB Meth-anol FC Hybrid	AB Hydro-gen FC Hybrid	AB Electric
Total running costs	0.056	0.053	0.047	0.049	0.044	0.049	0.056	0.050	0.054	0.045
Fuel ex tax (percent of total)	0.014 (5%)	0.012 (4%)	0.007 (2%) FT=0.009	0.009 (3%)	0.005 (1%) FT=0.006	0.010 (3%)	0.014 (4%)	0.010 (3%)	0.015 (4%)	0.007 (2%)
Fuel tax	0.006	0.005	0.004	0.004	0.003	0.003	0.006	0.004	0.003	0.002
Other (oil, tires, main-tenance)	0.036	0.036	0.036	0.036	0.036	0.036	0.036	0.036	0.036	0.036
Total fixed costs	0.250	0.268	0.281	0.292	0.304	0.297	0.317	0.315	0.303	0.363
Insurance	0.050	0.052	0.053	0.056	0.057	0.056	0.057	0.057	0.057	0.063
License, excise tax, registration	0.020	0.022	0.023	0.024	0.025	0.024	0.026	0.026	0.025	0.030
Capital costs	0.180	0.194	0.205	0.212	0.222	0.217	0.234	0.232	0.221	0.270
Total costs	0.306	0.321	0.328	0.341	0.348	0.346	0.373	0.365	0.357	0.408

NOTE: AB = advanced body; FC = fuel cell; CNG = compressed natural gas; FT = Fischer-Tropsch.

SOURCE: Weiss et al. (2000).

costs (license, registration, and insurance) and on 20 percent per year of the new vehicle cost. Running costs are based on 20,000 km per year of travel. Fees for license and registration of $0.02 (RMB0.17) per kilometer (scaled by new vehicle cost relative to the baseline to represent some excise tax and other costs) are incorporated into the calculations, as well as insurance costs of $0.05 (RMB0.4) per kilometer, with half of the cost scaled by the purchase price. These assumptions are consistent with current U.S. analysis (Davis, 1999).

The annual operating costs reflect the assumption that the new cars would be sold in the United States in 2020. As cars age, their capital value decreases, and fuel and maintenance costs become a larger fraction of the decreasing total annual operating cost. Likewise, in countries where certain fuels are heavily taxed, the ratio of capital to running costs would be less for vehicles using those more expensive fuels. All these costs are subject to uncertainties inherent in the assumptions made in this analysis. The cost difference between the baseline vehicle and the most expensive option is 22 percent.

Like today, the total annual cost for a new U.S. vehicle in 2020 is dominated by the capital cost, which is tied to the vehicle cost. Estimates indicate that the more efficient vehicles from an energy consumption standpoint are the more expensive ones, and the charges associated with increased price more than offset any fuel savings at current U.S. tax rates. For fuel cell vehicles, the total operating costs vary from the baseline of about 0.30 per kilometer to about $0.37 (RMB2.5–3.1) per kilometer. This difference reflects the roughly 30 percent higher estimated purchase price for the fuel cell vehicles. The $0.41 (RMB3.4) per kilometer cost of the electric vehicle is mostly attributable to the increased capital cost associated with the storage batteries. Overall, only *large* differences in fuel costs or fuel taxes are likely to have a significant influence on the annual operating costs of new cars. For example, at a UK tax rate of $3.53 per gallon of gasoline (8.8 times higher than the U.S. rate), the baseline vehicle fuel tax would increase to $0.044 (RMB0.36) per kilometer and the total new baseline vehicle operating cost would rise to $0.343 (RMB2.80), about 13 percent higher than in the United States.

Overall Life Cycle Comparisons

The following general comments are based on the cases evaluated:

- Reducing vehicle weight improves life cycle efficiency. In the cases studied, a 10 percent weight reduction produced about a 10 percent reduction in energy but resulted in about a 10 percent increase in vehicle cost. A smaller, lighter "China car" could be considerably more efficient than the typical U.S.-size car studied.

- Hybrid technologies offer significant energy savings, particularly in urban driving cycles. Typically, these cars are more expensive because of the more complex drive trains. A partial hybrid that shuts the engine on and off in stop-and-go traffic can achieve some of the emissions reductions and energy savings of a true hybrid, with only a modest impact on cost.
- The diesel engine offers some improvements in efficiency at a somewhat higher cost, but challenges remain as to whether it can meet emerging emissions standards for nitrogen oxides and particulates.
- A hydrogen fuel cell car does appear to offer some advantages. It eliminates harmful vehicle emissions (although it can produce other emissions, depending on how the fuel is produced), but because of the cost and uncertainty about the development of the technology, it is not clearly a winner in the near term. Major investments in research and development (R&D) and in infrastructure will be required to move this technology into a significant market share over the next decade or two.

The uncertainty bounds allow for the possibility of revolutionary improvements in any of the technologies. Where these happen, a predominant new technology may emerge. Because the vehicle system has so many components, breakthroughs are possible in many different areas—perhaps in control systems, batteries, engine or fuel cell technology, or fuels.

VEHICLE COMPONENT TECHNOLOGIES

Vehicle Weight and Body

The dependence of fuel economy on vehicle weight is shown schematically in Figure 4-5 (Horton and Compton, 1984). In the past, the main incentive for substituting lighter materials has been to meet requirements for decreasing emissions below critical levels while minimizing the economic impact. In the United States, where fuel costs are not of major concern to vehicle purchasers at present, average vehicle weights have been increasing as sport-utility vehicle (SUV) sales increase. However, this finding masks the fact that vehicles of a given size are becoming increasingly efficient. Lighter materials, including high-strength steels, aluminum, and plastics, are replacing heavier carbon steels. The potential for weight reduction is considerable, but the present barrier is the higher costs of many of the lighter materials. Automakers are working hard on ways to bring down these costs.

The high-strength steels now being used extensively in new vehicles can provide, with minimal additional cost, equivalent strength at less

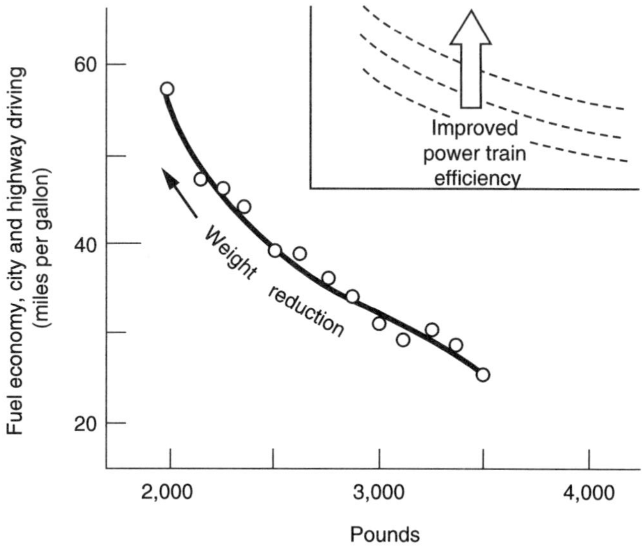

FIGURE 4-5 Typical fuel economy improvements in new vehicle options resulting from vehicle weight reduction for a typical power train efficiency. NOTE: Improvements in power train efficiency also increase fuel economy. SOURCE: Horton and Compton (1984).

weight. About a 15 percent reduction in vehicle fuel use can be achieved by substituting high-strength steel for regular steel. Additional weight reduction can be achieved through the increased use of aluminum or fiber-reinforced plastic composites. Substitution of these two light materials for steel must not, however, compromise vehicle safety. The costs and availability of virgin materials, as well as the costs of disposing of scrapped vehicles, will influence decisions about the level of recycling that is appropriate. Significant energy savings and some cost savings are associated with the recycling of metals, especially aluminum. However, aluminum alloys vary in properties as a function of composition, and so a mixed recycled alloy is unlikely to match the desired physical properties for particular car components. The car market is therefore still largely dependent on virgin aluminum. Some research is under way on developing an all-purpose alloy that would meet structural needs and yet be suitable for recycling. Other approaches are to design for a few alloys that would separate easily on disassembly. This mixing problem occurs with other materials as well and should be considered in the design for recycling.

Several European countries and Japan are enacting strict requirements for recycling or for manufacturers to be responsible for vehicles at the end of their useful life. A large vehicle fleet generates a considerable waste stream. In countries such as the United States where land is available for inexpensive disposal sites, this problem does not appear to be a priority one. But in congested countries where landfill disposal is not available and more expensive incineration is used for wastes, more aggressive measures seem appropriate. For metals, recycling is usually cost-effective, but other materials are more difficult to separate and are associated with only a marginal cost incentive (or a cost penalty) for recycling. Cars traditionally have not been designed for end-of-life waste minimization. However, some manufacturers are now modifying designs to facilitate disassembly, reuse of some components, and recycling of many materials. Mixed plastics and fiber materials can even be burned to generate process heat, for example. This new approach to design also has suggested ways in which repairs can be modularized for simpler servicing.

The Chinese government has examined domestic material resources for use in the manufacture of today's vehicles; where new materials are needed for the next generation of vehicles, similar analyses should be undertaken. Furthermore, the government will have to provide for the waste streams generated by scrapped vehicles, which it may wish to begin doing now in the requirements it imposes on industry—whether domestic or foreign.

Vehicle design comprises a complex set of decisions about aerodynamics, stability, interior space, and safety. The characteristics of a vehicle's propulsion system, transmission, and fuel storage systems are related to its weight and performance. Safety depends both on intrinsic performance capabilities and on crashworthiness. Today's new collision warning systems and other devices also may increase the safety of smaller vehicles. Vehicle design that seeks to protect the passengers in an accident requires a sophisticated combination of understanding the crush behavior of structural elements and the performance of passenger restraint and protection systems, such as air bags. It is certainly possible to design a very safe small car, but it still will be less safe than a larger car of similar design if it crashes into an object larger in mass.

Air drag and tire resistance provide opportunities for designers to reduce energy losses. Vehicle drag coefficients have fallen from the typical level of 0.3 in the 1990s to about 0.25 today—and even below for some of the PNGV concept cars (but at significant cost and with the elimination of outside mirrors and other items). Drag reduction is most important during high-speed driving conditions. Improved tires also are providing some benefits, such as traction, reduced noise, and ride comfort. But benefits can be lost if drivers do not maintain properly inflated tires. Cost is a factor, although improvements in tire efficiency affect safety as well.

Propulsion Systems

Conventional Propulsion Systems

Today's vehicles typically use gasoline or diesel engines as their primary power source. These engine configurations have been developed over the years to provide reliable and easy-to-operate sources of vehicular mechanical power. It is expected that these technologies will continue to evolve and improve over the next decade as well.

The gasoline engine serves most personal transportation vehicles worldwide. Gasoline internal combustion engines (ICEs) burn air-fuel mixtures using spark ignition (SI) to initiate combustion. They are capable of operating over a broad speed range, from several hundred revolutions per minute (rpm) to as high as 7,000 rpm, and of starting rapidly over ambient temperatures ranging from –35°C to well over 38°C.

Much of the improvement in engine efficiency over the last two decades in the United States has resulted indirectly from increasing the engine's specific power, measured in kilowatts (kW) per liter (rated power per liter of engine displacement). This achievement has enabled engine downsizing of some 58 percent and a 26 percent reduction in the average 0–60 miles per hour (mph) acceleration time (An et al., 2001a; DeCicco et al., 2001). The prospects for further increases in specific power are excellent (Jost, 2002; NRC, 2002), which translates into further engine downsizing while maintaining vehicle performance. Engine downsizing implies reduced engine friction and weight. Specific power has been increased by the addition of valves, fuel injection, improved fuel/air controls, low-friction and lightweight materials, higher engine speed, turbo charging, application of numerical analysis techniques to optimize engine processes, precision manufacturing, and greatly improved quality control. Today, most production gasoline engines rely on a homogeneous stoichiometric fuel/air mixture for the internal combustion process. The choice of this combustion process for production engines stems from its flexibility for operation over broad speed ranges, combined with catalyst exhaust after-treatment technology to control effectively vehicle exhaust emissions. Although there is little room to improve thermal efficiency significantly for this type of engine, some potential exists to improve part-load engine efficiency by reducing engine friction and pumping losses, resulting in an enlarged high-efficiency area on the engine performance map. This improvement can be achieved by reducing throttling losses through various valve train control technologies. Examples of such technologies include variable valve timing, variable valve lift, and throttleless "valvetronic" engine technology introduced by BMW in its 7-series sedan. Other technologies, such as cylinder deactivation and the ability to vary the compression ratio (see Flynn et al., 1999; Jost, 2002), also improve engine efficiency significantly.

Recently, many engine manufacturers have initiated efforts to develop gasoline direct-injection (GDI) stratified lean combustion processes to improve both engine thermal and part-load efficiency. Japanese manufacturers have introduced this technology to Asian markets. The GDI stratified lean-burn technology presents several new problems for engine designers. Similar to diesel direct-injection engines, the process is initiated with the injection of liquid fuel directly into the combustion chamber, forming regions of stratified rich and lean fuel/air ratios in the combustion space. The rich zones yield carbon-based particulates, which must be trapped with particulate filters. The lean zones yield nitrogen oxide (NO_x) emissions combined with available oxygen in the exhaust stream. This combination of NO_x emissions and available oxygen render three-way catalyst processes ineffective in removing nitrogen oxides; such systems require the use of new catalyst processes that effectively remove nitrogen oxides from an environment that includes free oxygen.

Diesel engines that have been developed for broad use in passenger and commercial vehicles operate over a somewhat narrower speed range than gasoline engines (Flynn et al., 1999). Today's diesel engines with high-quality fuel systems operate at speeds of between 500 and 4,000 rpm. In these engines, which operate at higher pressures than spark ignition engines, cylinder combustion is initiated by injecting fuel into high-temperature compressed air, causing compression ignition (CI) (Naber and Siebers, 1996; Dec, 1997; Siebers, 1999). Compared with gasoline engines, diesel engines are more difficult to start rapidly under cold ambient conditions. The minimum starting temperature for diesel engines without special starting aids is typically 0°C. Although diesel engines were once prone to produce more noise and vibration than gasoline engines similar in size, recent design developments have produced smooth running, quiet diesel engine configurations that are barely distinguishable from gasoline engines (Flynn, 2000).

New diesel engines incorporate a wide variety of technologies that improve performance and fuel economy and reduce emissions. Most new diesel engines apply high-injection pressure, which is enabled by a common-rail unit injection system with advanced injection timing management, turbocharging, aftercooling, and an integrated exhaust gas recirculation (EGR) manifold system.

Diesel cars have significantly penetrated markets in Europe and elsewhere, but future emissions standards are likely to challenge the ability of diesels to meet NO_x requirements (this is discussed more specifically later in the section Diesel Engine Emissions). At a somewhat higher cost than spark ignition engines, diesels offer improved efficiency and are the technology of choice for hauling heavy loads in freight transport where the fuel savings outweighs the initial capital cost investment.

Figure 4-6 illustrates the performance differences between the three types of propulsion technologies (fuel cell, diesel, and spark ignition) for a vehicle similar to the Volkswagen Golf, using the European driving cycle and based on the present state of technology (Wengel and Schirrmeister, 2000). The potential for improvement exists in each of the technologies. The figure is based on a prototype fuel cell design utilizing hydrogen. Fuel cell propulsion offers advantages in efficiency, especially for low-speed operation and for idling conditions in which the fuel cell output goes to charge batteries or in which the system is shut off (hybrid vehicles also offer similar advantages that are described later in this chapter).

The efficiency curves shown in Figure 4-6 are for a particular vehicle configuration and driving cycle. The relationship between performance and efficiency under different driving conditions is much more complex. Vehicle designers use "performance maps," a plot of normalized torque—expressed as brake mean effective pressure (BMEP)—from the engine as a

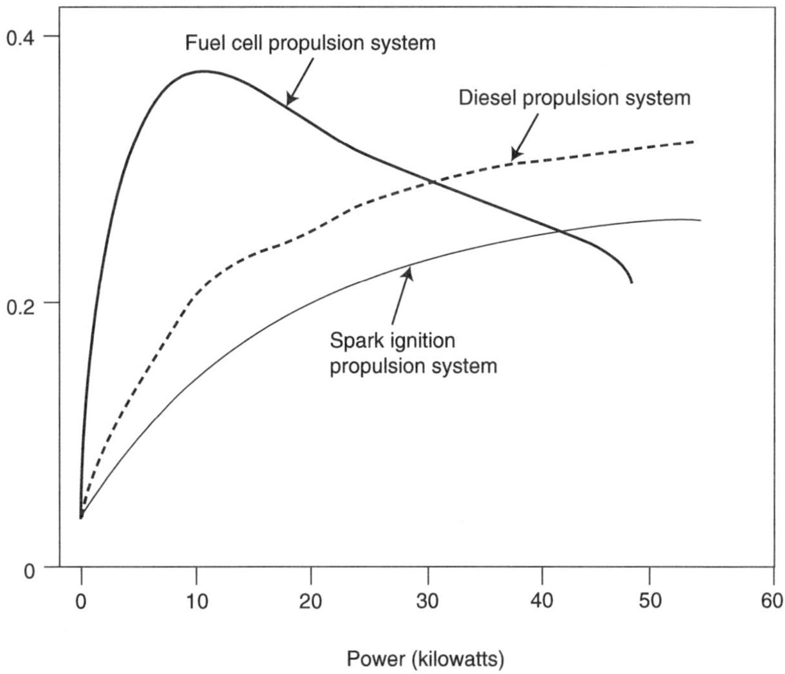

FIGURE 4-6 Comparisons of power train efficiency of combustion engine and fuel cell systems. NOTE: Information based on car similar to a Volkwagen Golf. SOURCE: Wengel and Schirrmeister (2000).

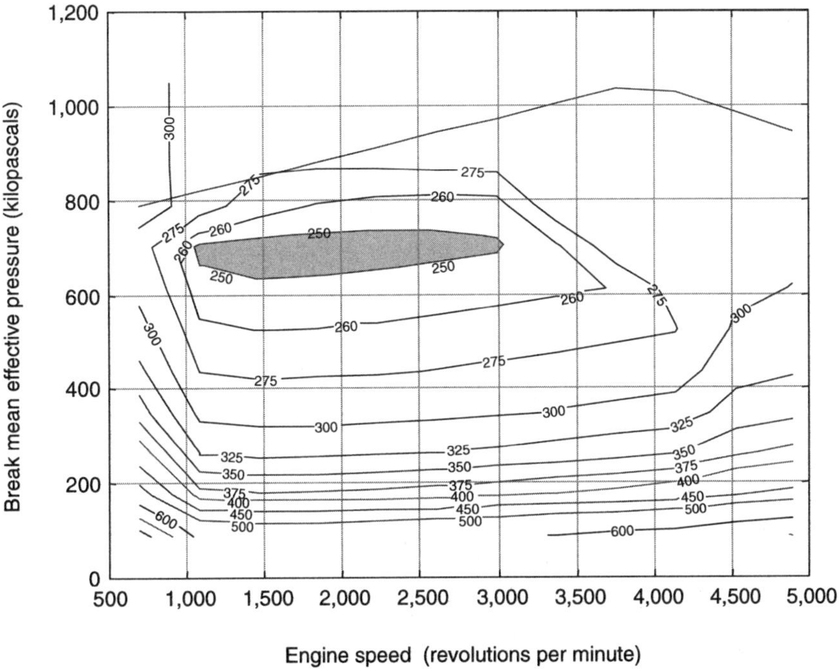

FIGURE 4-7 Typical performance map for a spark ignition engine. NOTE: Brake-specific fuel consumption (BSFC) contours are shown in grams per kilowatt-hour. Performance map based on a 3-liter, 155-brake horsepower (bhp) engine. SOURCE: An et al. (2001a).

function of engine speed in revolutions per minute. Figure 4-7 shows a typical performance map for a spark ignition engine.

Performance maps depict how efficiency varies throughout the range of operation associated with particular driving cycles, with various patterns of requirements for power and acceleration. A shaded region of "maximum efficiency" is indicated in Figure 4-7 (contours are indicated in grams of fuel per kilowatt-hour). As engine torque and speed combinations move outside this region, fuel efficiency drops.

The characteristics of the combustion engine are closely linked to the design of the transmission system to provide the desired performance characteristics for the vehicle. Trade-offs between performance and efficiency occur throughout the driving cycles. For U.S. cars, where fuel economy has been of less concern to the consumer than performance, the trade-off usually favors performance. European and Asian car manufacturers, who are seeking sales where fuel efficiency is valued more than high-speed acceleration and other performance factors, make a different trade-off in their designs. In the future, however, it appears the stronger emphasis will be on fuel economy because of concerns about foreign oil dependence and GHG emissions.

Emerging Propulsion Technologies

The homogeneous charge compression ignition (HCCI) diesel engine differs from both the typical gasoline and diesel combustion processes in that the energy release does not take place in a flame front. Fuel oxidation is accomplished by inducting a lean premixed charge of fuel and air into the engine cylinder. The lean premixed charge is subsequently compressed to higher temperatures and pressures until reactions similar to those encountered in gasoline engine knock phenomena occur (Westbrook et al., 1991). These reactions take place at temperatures considerably lower than those occurring in flame propagation, but still can be completed in the time allowed for piston engine combustion. If the combination of fuel/air ratio and starting temperatures and pressures are controlled so that peak combustion temperatures do not exceed 1850 K, the combustion process can proceed to completion without the formation of any particulate or NO_x emissions. The temperatures and pressures at which such processes take place are determined by the ignition characteristics of the fuel used. If high-octane fuels such as natural gas are used as the primary fuel for HCCI systems, the indicated efficiency of the process can approach that of the diesel engine. If low-octane fuels such as diesel are used, the pressures and temperatures at which these knock-like reactions take place are too low to allow compression and expansion ratios of above 8:1. Because these expansion ratios are similar to those in gasoline engines, the fuel economy of the cycle using diesel as its main fuel would be similar to that in gasoline engine operation, but without the output of the usual pollutants. Power output also would be low because of the very lean fuel/air ratio.

Researchers are exploring the HCCI ignition process for use in both light-duty vehicular applications and heavy-duty engine applications. Research presently in the feasibility study phase indicates that significant improvements in light-load efficiency can be achieved when using this process in light-duty gasoline engines. The process is very difficult to control, however, because it depends on reaching specific combinations of pressure and temperature for its initiation. The lack of a spark or injection event to control the initiation of combustion makes it more difficult to coordinate variables such as intake temperature, fuel octane number, or intake pressure, which must be controlled on a cycle-by-cycle basis to manage the process. Because small variations in temperature cause large differences in the times of combustion within the cylinder, it is unclear whether intake processes can be managed closely enough to control HCCI combustion. Presently, funding of research on the HCCI combustion process is quite large. But it remains to be seen whether viable systems can be developed to incorporate such an approach into the production engine.

Electric propulsion for cars has generated great interest because of its potential for "zero emissions" during use. Emissions may be generated in the production of the electricity, but when the generating plant is located outside of the urban air shed, such emissions may be of less concern to those residing in polluted urban areas, and yet they may be transported hundred of miles to affect the air quality of others. Electric power generation from fossil fuels does produce GHG emissions, however.

The main barrier to the use of electric energy in transportation vehicles is the difficulty in storing electricity. At present, batteries are heavy, cumbersome, and expensive. And charging times are long relative to liquid fueling times for a vehicle. Researchers are seeking a more efficient battery storage system, but a major breakthrough is needed if electric cars are to compete in price, convenience, and range with today's liquid fuel vehicles. Although applications for electric vehicles do exist, especially for short travel distances, they are unlikely to be a major competitor with conventional vehicle technologies in the next decade or two.

Fuel cell propulsion systems offer an alternative way to produce electricity for propulsion from onboard fuels. Most fuel cells, especially those for transportation applications, operate with hydrogen fuel that can be either stored on board or chemically reformed from gasoline or other liquid hydrocarbon fuels. Liquid methanol can be used as a direct fuel for a fuel cell, but the technology is still far behind the hydrogen fuel cell technology. A fuel cell is an electrochemical device that produces electricity by separating the hydrogen fuel into electrons and protons (hydrogen ions) via a catalyst. Because the fuel is converted directly to electricity, a fuel cell can operate at higher efficiencies than internal combustion engines, extracting more electricity from the same amount of fuel. The fuel cell itself has no moving parts, making it a quiet, reliable source of power.

Fuel cell technologies are presently being developed for a variety of applications (see the appendix to this chapter for a brief description of the alternatives). The most promising fuel cell technology choice for transportation applications is the proton exchange membrane (PEM) fuel cell. These cells operate at relatively low temperatures (about 200°F or 95°C) and have high power density. They can vary their output quickly to meet shifts in power demand and are suited for applications—such as in automobiles—where quick start-up is required. PEM fuel cells are the primary candidates for light-duty vehicles, for buildings, and potentially for much smaller applications such as replacements for rechargeable batteries. The proton exchange membrane is a thin plastic sheet that allows passage of hydrogen ions. The membrane is coated on both sides with highly dispersed metal alloy particles (mostly platinum) that are active catalysts. Hydrogen is fed to the anode side of the fuel cell where the catalyst encourages the hydrogen atoms to release electrons and become hydrogen ions (protons). The electrons travel in the form of an electric current that can be utilized before it returns to the cathode side of the fuel cell where oxygen has been fed. At the same time, the protons diffuse through the membrane to the cathode, where the hydrogen atom is recombined and reacted with oxygen from the air to produce water, thus completing the overall process.

Comparing Fuel Cell Systems with Gasoline and Diesel Engine Systems

Figure 4-6 compared the typical system efficiencies of a gasoline internal combustion engine, a diesel engine, and a fuel cell for current technologies. Although both gasoline and diesel engines have very low part-load efficiencies, the efficiency of a fuel cell system peaks around 20 percent of full load. Figure 4-8 shows how typical gasoline engine efficiency varies with percentage of maximum engine power, along with the operating modes of a U.S. car in an urban driving environment. For U.S. driving behaviors, average engine power demands occur at about 10 percent of maximum engine power. This level is well below peak efficiency for the gasoline and diesel engines, but it is where the fuel cell is most efficient. Thus potentially a fuel cell vehicle can be much more efficient than its internal combustion engine counterparts.

Fuel cell vehicle developers must overcome many technological and economic challenges if they hope to match the performance and cost of today's conventional technology vehicles. Fundamental problems with fuel cell technology are fuel selection, generation, distribution, and storage. The only truly zero emissions vehicle fuel cell is the direct hydrogen fuel cell. However, hydrogen infrastructure and onboard storage pose a huge challenge. Gasoline infrastructure and onboard storage to provide fuel for fuel cells already exist, but unfortunately a reformer that can convert gasoline to hydrogen adds more weight and technical complexity to a car, and in situ reforming technology is still far off for lower-temperature automotive fuel cell technologies. Furthermore, "reforming" liquid fuels to make hydrogen still generates GHG emissions and deteriorates vehicle start-up and transient performance.

For direct hydrogen fuel cell vehicles, one critical technological issue is onboard hydrogen storage. Hydrogen can be stored on board vehicles in many forms, including as compressed gas, as liquid, or within metal hydride alloys. Table 4-3 compares some competing hydrogen storage technologies and compares those technologies with gasoline and other

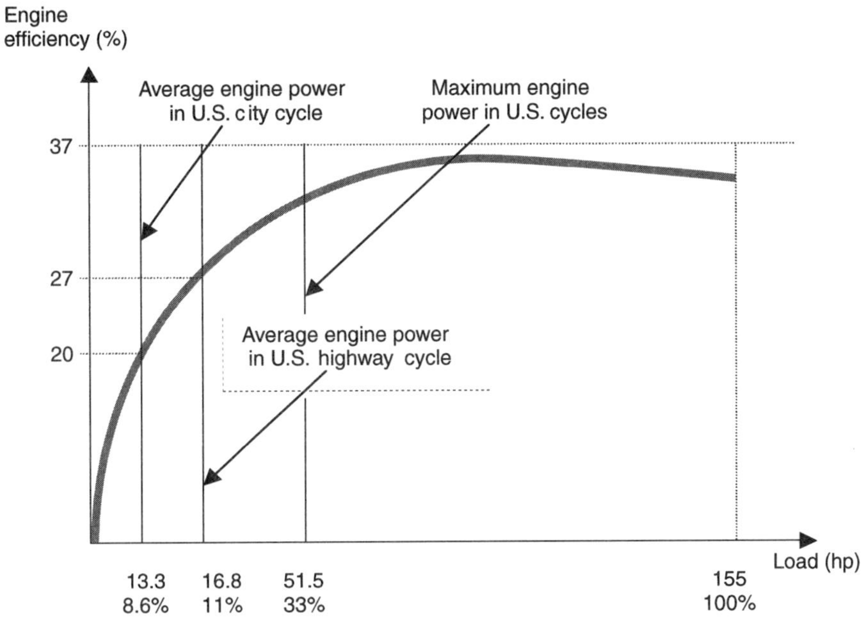

FIGURE 4-8 Typical engine efficiency and average driving cycle operating modes for U.S. cars. Basis: Approximate efficiency versus horsepower (hp) curve for a 155 hp engine. SOURCE: Calculations by Feng An.

energy storage media. Both the volume and weight of the hydrogen storage are based on the equivalent performance of 15 gal of gasoline fuel used in a typical U.S. car giving a 400-mile range. It is estimated that about 6 kg of hydrogen are needed to drive a fuel cell vehicle (that is smaller and more efficient) for 400 miles (Ashley, 2001). The comparison in Table 4-3 assumes that the hydrogen fuel cell vehicles are about twice as efficient as diesel vehicles.

The last four columns of Table 4-3 show simply that, compared with gasoline technology, today's hydrogen storage technology faces challenges in both volumetric efficiency and weight penalties. Furthermore, a hydrogen infrastructure would have to be developed and its costs would be affected by the low density of hydrogen, which would require high-pressure transmission and distribution lines. Because hydrogen also has very wide flammability limits, safety will be a concern, especially during fueling because consumers are accustomed to fueling cars with liquid fuels. Overall, the infrastructure costs will likely run into the billions of dollars.

Beyond technological challenges, the costs associated with fuel cell

TABLE 4-3 Typical Energy Content and Storage Requirements for Automotive Energy Sources

Energy Type	MJ/kg (LHV)	MJ/liter (LHV)	Percent of Gasoline Energy Density	Volume, Liters	Volume, Gallons	Weight, Kilograms	Weight, Pounds
				Fuel Storage Quantities Equivalent to 15 Gal Gasoline or 1,800 MJ (ex tank)			
Gasoline	32.0	43.7	100	56.4	15.0	40.9	90
Diesel	36.0	41.8	112	50.1	13.3	42.8	94
Methanol	16.0	20.1	50	112.8	30.0	88.9	195
Methane (STP)	0.036	50.0	0.11	50,133	13,300	35.7	78.6
				Fuel Storage Quantities Equivalent to 6 Kg of Hydrogen or 750 MJ (ex tank)			
Hydrogen (STP)	0.011	120	0.03	66,700	Gaseous	6	13
Hydrogen (gas, 350 atm)	3.25	120	9.3	256	80.2	120	267
Hydrogen (liquid, 20 K, 1 atm)	5.05	72	14.4	82.7	36.7	28	61
Metal hydride	11.9	1.4	34	60	16	500	1,100
Mg-based metal hydride (ECD)	6	2.4	17	120	32	300	660
NiMH battery		0.25					
USABC comm. goal 2000		0.54					

NOTE: Energy equivalent to 15 gallons (gal) of gasoline; hydrogen based on 6 kilograms (kg) of fuel stored. LHV = lower heating value; STP = at standard temperature and pressure; ECD = Energy Conversion Devices Inc.; Mg = magnesium; NiMH = nickel metal hydride; USABC = U.S. Advanced Battery Consortium; MJ = megajoule; atm = atmosphere.

SOURCES: Stodolsky et al. (1999); Wang (1999); Wang and Huang (1999).

systems, including fuel cell stacks, system accessories, and the onboard reformer, remain a major barrier to the commercialization of fuel cell technology. Today, gasoline power trains cost around $25 (RMB207) per kilowatt, diesel power trains about $50 (RMB415) per kilowatt, and limited production PEM fuel cells (e.g., the Ballard Model Mark 900), about $500 (RMB4,100) per kilowatt (Ashley, 2001). Fuel cell costs are expected to decrease in the future as technology advances and production grows. The U.S. Department of Energy (2001) estimates that if PEM fuel cells had been mass-produced (500,000 units per year), the cost would have been about $200 (RMB1,700) per kilowatt in 2001 and could be reduced to $125 (RMB1,000) per kilowatt by 2005—the technical target for the proposed U.S. FreedomCAR (Cooperative Automotive Research).

Emissions Control Systems

Gasoline Spark Ignition Emissions

For gasoline internal combustion engines, three-way catalyst systems are highly effective (99+ percent) in achieving Euro II emissions standards. Figure 4-9 portrays the engine out and tailpipe out emissions of a typical three-way catalyst-equipped spark ignition engine. The catalysts, which are composed of mixtures of noble metals supported on a ceramic substrate, are used to eliminate carbon monoxide, nitrogen oxides, and hydrocarbons from gasoline engine exhaust. But they are expensive and can be poisoned by impurities in the fuel—notably lead, which is now being phased out of gasoline in China. The effective operation of these catalyst systems depends on maintaining a stoichiometric fuel/air ratio in the engine's combustion chamber. This stoichiometric operation provides the catalyst with an oxygen-depleted exhaust stream containing an appropriate level of hydrocarbon emissions so that, in the presence of the catalyst, the reducing atmosphere liberates oxygen from the nitrogen oxides and supplies them to the unburned hydrocarbons for oxidation.

The catalyst combinations required to complete the joint reduction and oxidation processes have been developed after many years of empirical work by engine manufacturers and catalyst suppliers. Because a three-way catalyst is not efficient under cold operating conditions, recent work has focused on reducing tailpipe emissions during a vehicle cold-start, which contributes more than 75 percent of the emissions of modern U.S. cars. Solving the problem requires development of quick "light-off" systems. Cooperation between the various suppliers of the three-way catalysts and the manufacturers of their associated electronic control systems is needed to manage performance during start-up and other transient conditions. Although Chinese industry can supply these components, it has

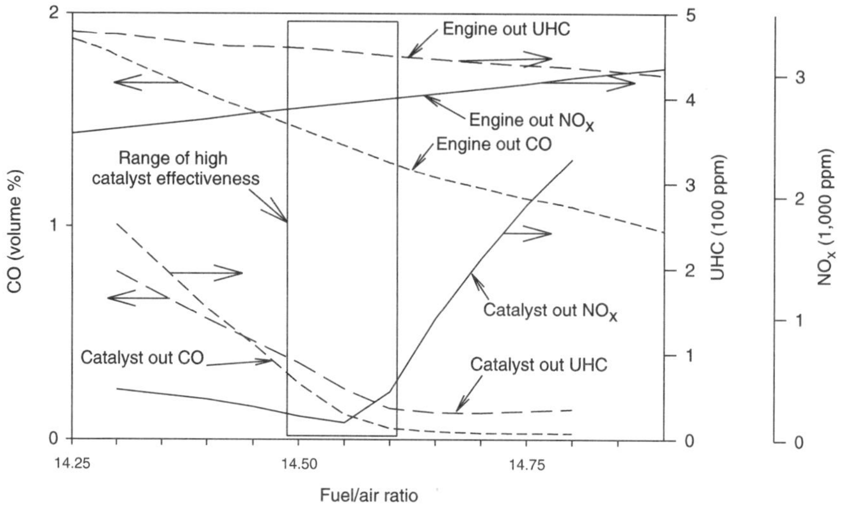

FIGURE 4-9 Three-way catalyst system behavior versus fuel/air ratio. NOTE: UHC = unburned hydrocarbons; NO_x = nitrogen oxides; CO = carbon monoxide; ppm = parts per million. SOURCES: Data replotted from Sher (1998: Figure 6.4); catalyst effectiveness: Kummer (1981).

not yet developed an independent capability for designing and optimizing overall emissions control systems.

Meanwhile, a promising new catalyst technology under development offers much the same efficiency at a substantially lower cost by replacing some of the noble metals currently used with rare earth elements. Use of these less expensive catalysts depends on reducing the sulfur in the fuel to levels significantly below those currently being planned (Zhan et al., 2001).

Diesel Engine Emissions

The diesel combustion process yields emissions of particulates and nitrogen oxides. Just as for the gasoline engine, meeting the most rigorous of the emissions standards proposed today will require applying exhaust aftertreatment devices to diesel engines to bring their emissions of nitrogen oxides, unburned hydrocarbons, and particulates to the desired low levels. Because the exhaust of a diesel engine is cooler and contains more oxygen than that of a gasoline engine operating under stoichiometric conditions, removing the pollutants from the exhaust of a diesel requires a somewhat different technology. The particulates in the diesel exhaust pose an additional problem.

The NO_x emissions from diesel engines can be controlled by cooling the diffusion flames within the engine in one of a variety of ways (De Witt and Wan, 2000). The most common is to add diluents, in the form of water-fuel emulsions, inert gases, or recirculated cooled exhaust gas, to the combustion process. All these methods provide additional thermal mass near the diffusion flame and thus limit the overall rise in the flame temperature. The addition of diluents (exhaust gas recirculation or EGR) is the most effective way to control in-cylinder diesel NO_x emissions.

As shown in Figure 4-10, the lowest NO_x emissions level from diesel engine combustion is 5.5 g per kilogram of fuel burned for engines operating at 1,500 rpm (Flynn et al., 2000). These results at 1,500 rpm translate to lower numbers at higher engine speeds, because NO_x conversion is directly proportional to combustion residence times. These levels of NO_x production are significantly above the minimum NO_x levels produced by gasoline engines operating with effective three-way catalyst systems. Therefore meeting newly legislated NO_x emissions targets will require additional aftertreatment of the exhaust gas stream with selective catalytic reduction (SCR) or other techniques to remove nitrogen oxides.

Diesel engines also have particulate emissions. The rich combustion process converts a significant portion of the carbon mass in the fuel to particulate precursors. Those particulate precursors that avoid going through the vigorous diffusion flame are left as tailpipe emissions. The California Air Resources Board (CARB) has labeled diesel particulates as toxic emissions, and such emissions have been associated with degradation in lung function by a variety of epidemiological studies. Particulates also degrade visibility as they accumulate in the atmosphere. As such, to satisfy future particulate emission legislation in the United States the manufacturers of diesel engines probably will have to add exhaust particulate filters to their products. These filters must trap the particulates and then, by managing temperatures, provide an opportunity for the oxidation of the particulates on the trap. Under favorable operational and ambient conditions, this combination of trapping and oxidation can be completed without additional management or manipulation of trap temperatures. If ambient conditions or operational constraints prevent exhaust temperatures (and thus trap temperatures) from rising to the level needed to oxidize the particulates, trap temperatures may require active management through the injection of additional fuel or the modification of engine operation to raise exhaust temperatures.

Fuel Cell Emissions

A fuel cell operating on hydrogen emits only water as a waste stream. However, when fuel cells operate through reforming a fuel such as gaso-

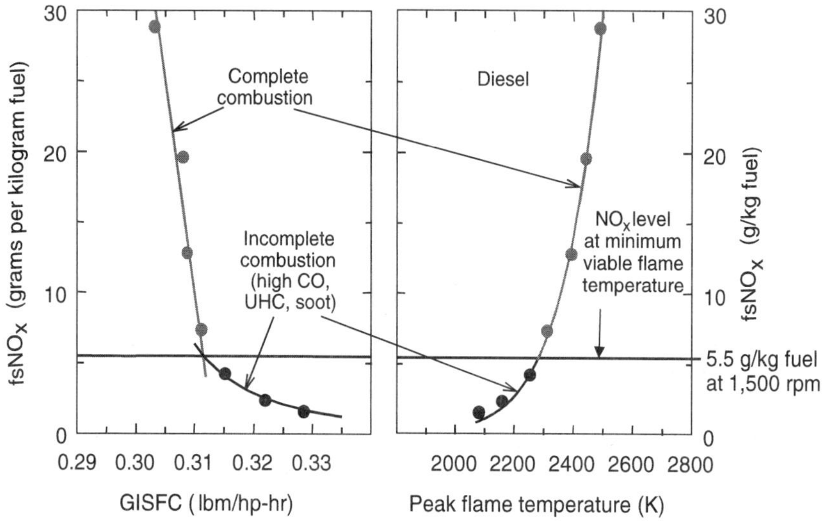

FIGURE 4-10 Limits of diesel combustion at 1,500 revolutions per minute (rpm). NOTE: GISFC is the fuel consumption per net power produced in the compression and expansion strokes of engine operation and thus disregards the gas exchange strokes of engine operation; fsNO_x is the NO_x emissions on a fuel-specific basis. EGR = exhaust gas recirculation; UHC = unburned hydrocarbons; NO_x = nitrogen oxides; g/kg = grams per kilogram; lbm/hp-hr = pounds mass per horsepower-hour. SOURCE: Flynn et al. (2000).

line to produce hydrogen, GHG emissions may be produced by the reformer. Because the reformer operates at a lower temperature than a combustion engine, NO_x emissions are negligible. A clean fuel (low sulfur) is required to maintain satisfactory performance of the system. This higher-efficiency propulsion system also provides a benefit in reducing emissions per kilometer.

Meeting Future Emissions Standards

For diesel engines, both the particulate and NO_x aftertreatment systems are likely to require active management of the exhaust system temperature and the fuel/air ratio. Presently, such systems are in their earliest demonstration phases, and much work remains to determine whether they can be produced for a wide range of operating requirements and environments. Figure 4-11 presents the European, Japanese, and U.S. emissions standards, along with the demonstrated capabilities of typical diesel and spark ignition systems. NO_x emission levels are depicted in grams per kilogram of fuel burned, a unit that serves to represent the level of technology required for the removal of nitrogen oxides. Because emissions standards are set on a grams-per-mile or grams-per-kilometer basis, as fuel mileage improves, more stringent emissions standards can be achieved. Thus, as shown in Figure 4-11, a car achieving 20 miles per gallon (mpg) might have difficulty meeting the Euro III diesel standard, but if the mileage for that diesel engine were improved to 40 mpg, it could clearly meet the proposed Euro III diesel standard. Figure 4-11 indicates that some level of exhaust aftertreatment would be required to remove nitrogen oxides for diesel engines as the emissions limits are lowered. With three-way catalyst technology, the gasoline engine can achieve or further reduce emissions as required by the U.S. 2007 light-duty standard.

The diesel engine will benefit from many of the possible improvements described earlier for the spark ignition engine—for example, vehicle weight reduction, improved component efficiencies, recovery of kinetic energy during braking, and engine turn-off when stopped. Vehicle hybridization is one way in which to realize the latter two improvements (see the section Hybrid Vehicle Technologies).

The advantages of the diesel engine for hauling heavy loads will continue to make it the leading option for trucks. Now that the pollution produced by trucking is gaining more attention, new emissions standards are being applied to heavy-duty vehicles. The technologies developed for improving the emissions performance of heavy-duty vehicles also may become available for lighter vehicles in the future.

Transmissions

Matching the torque and speed requirements of a vehicle's drive wheels with that of the engine's capability requires a transmission that permits a variety of operational gear ratios. Such transmissions can be shifted manually, electromechanically, or automatically from one gear ratio to the next. Mechanical transmissions usually have the highest overall transmission efficiencies, typically between 90 and 97 percent. In recent years, hydrodynamic transmission drives have been augmented with torque converter lockup mechanisms so that the hydrodynamic losses can be minimized when the overall transmission gear ratio is 1:1. Most vehicles in the United States use hydrodynamic automatic transmissions.

Substantial improvements in transmission performance are under way (DeCicco et al., 2001). The main developments are the addition of extra gear ratios to conventional transmissions (five- and six-speed automatics), motor-driven gear ratio shifting (which allows smart electronic control and eliminates the torque converter—a source of friction losses, especially in urban driving), and continuously variable transmissions.

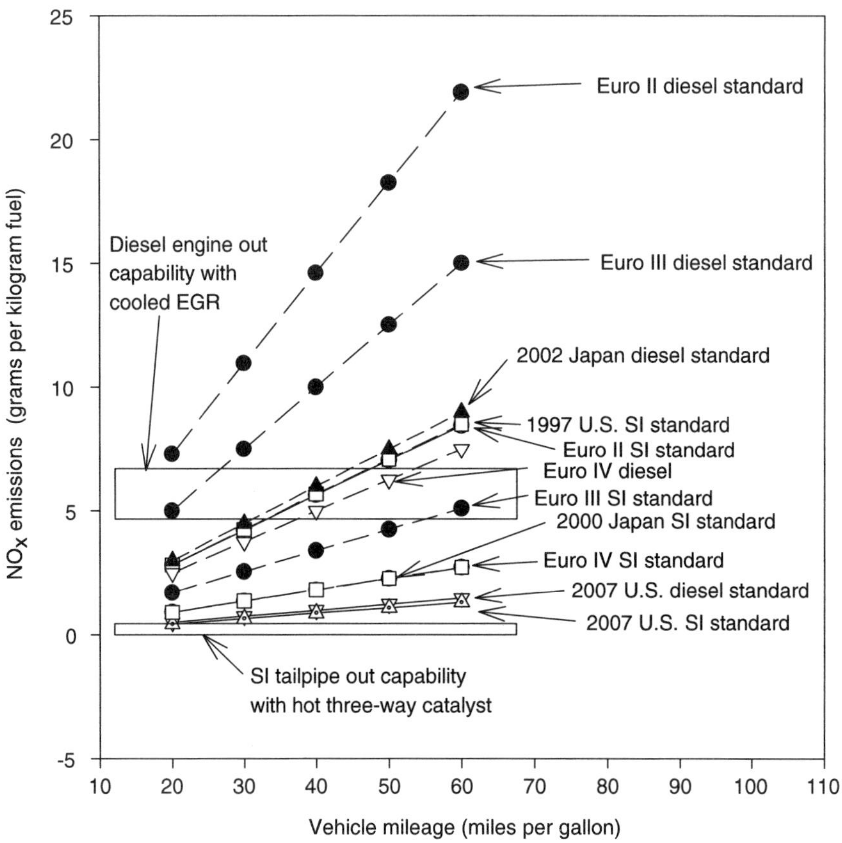

FIGURE 4-11 Fuel-specific nitrogen oxide (NO_x) emissions standards versus capability. NOTE: SI = spark ignition; EGR = exhaust gas recirculation. Fuel-specific gravity gasoline = 0.75; diesel fuel = 0.85. SOURCES: Standards: Table 4-1; capabilities: Flynn (2001).

Added gear ratios allow the engine to turn at modest speeds (near the maximum efficiency zone shown in Figure 4-7) over a range of vehicle speed and acceleration conditions. The continuously variable transmission uses more complex technology to approach an infinite number of gear ratios in order to optimize engine speed over variable driving conditions and to permit the engine to always operate in the maximum efficiency zones. The cost implications of these advanced transmission systems are uncertain, but there appear to be enough potential improvement possibilities so that the costs of evolved transmission systems will not have a major impact on vehicle costs.

Some European and Asian vehicle manufacturers have offered production versions of continuously variable transmissions. Such devices typically use tapered belts and pulleys to change effective gear ratios so that the engine is always operating at maximum efficiency. Torque is transmitted through frictional forces at the belt pulley interface, and speed is varied by changing the relative diameter of the input and output pulleys. Today such devices are being offered only on very small vehicles because of the limited torque-carrying capability of such frictional drives.

Electrical Systems

In the early days of automobiles, batteries were used first for lights and then, after the invention of the electric starter in 1912, to start the engine automatically. The generator was developed concurrently. The 6-volt (V) dry cell or lead acid battery was the standard in the first half of the twentieth century. As cars evolved, the 6 V systems proved inadequate, and in the 1950s the industry changed to a 12 V lead acid battery. Since then, the addition of control systems and more electrical amenities in vehicles has steadily increased the electrical power requirement for cars, but large current draws cause voltage drops and significant parasitic power losses in today's electrical systems. It is estimated that a 12 V system can support only up to 3.5 kW electrical loads. The present average power demand in an automobile is about 1.2 kW. Emerging technologies that will require even greater electrification are discussed later in this chapter.

The industry is considering a transitional dual voltage system, though the car of the future is likely to use a single 36 V (42 V output generator rating) power distribution bus with provision for hybrid operations (direct current and alternating current) and intelligent controls with multivoltage power distribution and load management. Some current production vehicles already are using 42 V machines for the starter-generator. Although these new systems are likely to be introduced first in luxury vehicles because of cost, they also have significant performance advantages for smaller cars. Transition to the higher voltage could reduce both power losses and the weight of wiring harnesses. However, advanced power control and distribution systems will be needed to operate the more complex vehicles of the future.

For vehicles that use electric motors for propulsion, such motors could, in addition to powering accessories, serve as the starter-generator and support regenerative braking, producing higher efficiencies over a wider speed range. The traction motors, however, are still in the developmental stage. Permanent magnet motors require costly materials and are limited in operational speed at 42 V by back emf (back voltage) problems.

Induction motors tend to overheat and thus lose efficiency at 42 V (it is hard to cool a moving rotor). Both induction and permanent magnet motors could benefit from an even higher voltage system, but that creates other problems, including ones of safety. Switched reluctance motors are under development to avoid the problems of back emf and overheating, but the present generation of motors has noise and vibration problems. Further research and development, however, will likely uncover solutions to all these challenges, opening the possibility that over the next decade the rate of introduction of 42 V automotive technologies into passenger cars will increase substantially (Ehsani et al., 2001).

Electronic Controls

Because they provide the increased sophistication needed to meet higher emissions standards while providing good fuel economy and good vehicle drivability, electronic controls have become ubiquitous in modern automobile technology. In fact, in recent years these controls have evolved to digital systems that are fully programmable. At the current level of technology, a programmable digital computer is coupled to advanced sensors that provide real-time data to allow the inference of various engine operating parameters such as specific engine pollutants, fuel consumption, engine horsepower, and engine torque. Using these inputs, the computer instantaneously controls spark, fuel delivery, quantity of exhaust gas recirculated to the engine, and, in some cases, the transmission. Ambient parameters such as temperature and atmospheric pressure also may be measured.

The computer industry can easily provide the hardware to accomplish these tasks, but the bigger challenge is developing the control strategy that will optimize vehicle performance under a wide variety of ambient and driving conditions. Doing so will involve utilizing complex analytical models that relate emissions, fuel economy, and vehicle drivability to detailed operating parameters of the vehicle power train—for example, its engine, transmission, and driveline. These parameters depend on the specific driving conditions and vary over wide ranges between engine idle and full power.

Usually, a limited amount of hardware-specific information is collected in the laboratory on various operating parameters, such as fuel consumption and emissions under specific operating conditions. These data are then fed into analytical models to provide the final control algorithms for each specific engine/vehicle combination. Because of the extraordinary effort devoted to creating the individual algorithms and because the control algorithms are of critical importance to the satisfactory operation of the vehicle, manufacturers consider both the details of the procedure and the algorithms to be highly proprietary. Algorithms for hybrid designs are significantly more complex and require additional levels of design skills and sophistication.

Emerging Capabilities

The growth in instrumentation and control capabilities in vehicles is leading to new possibilities for diagnosing vehicle problems and expediting repair—perhaps even while the vehicle is in use. Global positioning system (GPS) capabilities already are available to help drivers identify their location and navigate. In the future, these capabilities may extend to automated driving on specially equipped highways and other advanced technologies. Research is under way on automated highways, magnetically levitated vehicles, "drive-by-wire" systems, and other long-term technologies. However, most of these systems are at least a decade or more from realization and probably will require rethinking the traditional ways of delivering mobility to a diversity of consumers.

Fuels and Onboard Fuel Storage

Present Fuel Technologies

China's present road transportation sector is almost completely dependent on petroleum fuels, with some very limited use of electricity and compressed natural gas. Gasoline and diesel fuels are used widely in motorized vehicles because of their high energy storage density, compact and lightweight storage systems, and low cost relative to other fuels. Table 4-3 shows the energy storage volume and density values for various automotive fuels. Energy equivalent to fifteen gallons of gasoline (1800 MJ) is used as the basis for comparing alternative liquid fuels. For hydrogen, it is assumed that 6 kg of fuel is used (735 MJ) in a smaller, more efficient hydrogen fuel cell vehicle (Westbrook and Chase, 1988; Stodolsky et al., 1999; Wang, 1999; also see Chapter 5). The last four columns indicate the volume and weight of the fuel stored onboard. The last two rows in Table 4.3 show some battery data for comparison.

A liquid fuel storage tank in an automobile weighs less than the fuel it contains, whereas metal hydride storage systems, the heavy-walled pressure tanks required to store compressed gas, and the insulation systems needed for liquid hydrogen storage at very low temperatures can add significant weight penalties. Researchers are aiming for hydrogen storage systems that store about 5–10 percent hydrogen by weight and volume. Much research is under way on the storage possibilities of carbon fibers and other novel media, but significant progress is still needed to make a hydrogen storage system that does not impose too high a cost, volume, and weight penalty on a car.

The present petroleum fuels (domestic and imported) also will continue to change. To meet the Euro II emissions standards and the even more stringent ones to come, the *gasoline (from petroleum)* of the future will move toward very low sulfur content, with possible changes in volatility, aromatics, or other specifications driven by the increasingly stringent environmental emissions standards. *Diesel fuel (from petroleum)* will evolve from its current properties toward very low sulfur content, with possible changes in volatility, polynuclear aromatics, cetane, and other specifications

Some of the near-term alternative fuels might be feasible in China.

Natural gas is being used now as compressed natural gas in limited quantities for bus and taxi fleets in specific locations such as Beijing. CNG (primarily methane) is typically stored in vehicles as a pressurized gas (at about 200 atm pressure).

Methanol from coal or coal bed gas could be used as an additive to gasoline in an M15 (85 percent gasoline–15 percent methanol) mixture or as a pure fuel. Because methanol has a lower energy density than gasoline, it requires a larger fuel tank. Moreover, it is corrosive and so requires special metallurgy, and it is a contact poison and so must be handled with care. Finally, methanol from coal or other sources (such as natural gas) is generally more expensive than gasoline. Methanol fuel can be reformed on board to make hydrogen for a fuel cell. Overall, methanol from coal bed gas does not save net energy or reduce GHG emissions, but it could serve as a domestic fuel alternative for China.

Dimethyl ether (DME) from coal or coal bed gas offers a good alternative to gasoline and diesel. Wang (1999) estimates that DME, even if manufactured from remote gas, would be quite expensive, with a delivered (to the vehicle tank) cost (in 1995 U.S. dollars) ex tax of \$1.89 per gallon (RMB4.14 per liter) of gasoline equivalent. For comparison, diesel (50 percent Fischer-Tropsch) would cost \$0.65 per gallon (RMB1.42 per liter) and methanol \$1.20 per gallon (RMB2.63 per liter) when made from remote natural gas valued at about \$0.50 per gigajoule (GJ). Making DME from coal would cost considerably more than making it from remote natural gas. Moreover, manufacturing energy losses would be higher, and carbon dioxide (CO_2) emissions would be much higher. DME also is a pressurized gas at normal temperatures (boiling point of –24°C) and thus would require an entirely new and expensive fuel distribution system and new and costly changes in all vehicle fuel tanks and fuel systems. Vehicle range would be lower because of fuel storage limitations, especially for heavy-duty trucks.

Biofuels such as ethanol and methanol can be made from agricultural or other combustible wastes or from farmed energy crops such as corn, grain, or fast-growing cellulosic materials. Basically, biomass energy content is the result of a solar energy conversion process that operates at about 1–2 percent efficiency, so the energy content is relatively low per unit of planted area. Fossil fuels today are the result of eons of biomass conversion. Moreover, account must be taken of the costs and benefits of using land to produce fuel instead of food and of the pollution problems associated with producing biofuels.

Liquefied petroleum gas (LPG), mostly a mixture of propane and butane, has been used for years as a convenient "bottled gas" for remote locations, for camping, and for vehicle fuel. At 5–10 atmospheres (atm) pressure, saturated liquid can be stored at typical ambient temperatures. LPG is usually produced along with oil or natural gas, but it is not as plentiful.

If over the next two decades *hydrogen* achieves widespread use as a transportation fuel, it will likely be manufactured most economically by reforming natural gas at "service stations" located on a gas pipeline network. It probably will be stored and dispensed as a gas at about 350–400 atm pressure. Other, more expensive options include generating hydrogen from coal (Williams, 1998), electrolysis of water, or reforming natural gas in large, centralized facilities and piping compressed hydrogen or trucking liquid hydrogen to service stations. Any of these methods, however, will require large investments in infrastructure.

Although cars that utilize *electric power* are considered "zero emissions vehicles," emissions may still be associated with the original production of the electricity.

Although other technologies are being developed, nickel metal hydride (NiMH) batteries are the technology of choice for automotive applications today, both for hybrids and electric vehicles. Advanced lead acid batteries are less expensive than the NiMH batteries, but have a much shorter operational life.

Electric vehicle batteries currently have specific energy of about 70 watt-hours per kilogram (Wh/kg) and specific power of about 150 W/kg (U.S. Department of Energy, 1999; GMC, 2001). It is assumed that by the year 2020 electric vehicle battery performance will improve, especially the specific energy, and battery performance will be close to meeting the Advanced Battery Consortium's commercial goals of 150 Wh/kg and 300 W/kg (U.S. Council for Automotive Research, 2000). These commercial goals are judged to be the battery performance required to produce acceptable electric vehicle performance. Although the NiMH battery probably cannot reach this potential, another technology, such as the lithium-ion battery, may. Its specific energy is significantly higher than that of the NiMH battery technology. Batteries are not intended to be fully discharged, because such a step shortens their lifetime and decreases their capacity. Also, topping off a battery at a high state of charge is inefficient because of its internal resistance. Thus cycled battery applications tend to operate within a state of charge range of 20–80 percent.

For the entirely electric vehicle, both battery performance and charge density constraints (specific power and specific energy) are important. In addition to providing the power needed for peak motor power, the battery energy storage capacity must be sufficient to give adequate vehicle range. When a battery's specific energy is too low, the extra battery weight needed adds to the vehicle mass and thus requires additional structural support and increased motor power, generating an undesirable compounding effect. Given this constraint, the battery pack is selected based on its power capacity, and no effort is made to augment vehicle range beyond what the available electric vehicle battery technology can provide. The physical size of the battery also must be considered because of its possible intrusion into the vehicle's interior space.[2]

Although some of the fuel alternatives just described may have niche uses in China, a practical view suggests that petroleum fuels will be the main choice for the automobile fleet as it develops over the next few decades. Even for this choice, China will have to make significant investments in new fuel transportation and distribution infrastructure, including cleaning or replacing old facilities that are incompatible with the new, cleaner fuels.

HYBRID VEHICLE TECHNOLOGIES

Hybrid Power Trains

A typical combustion engine delivers peak efficiency only at a particular power level. Today's cars are designed with oversized engines that are able to provide peak power for acceleration in passing or climbing hills at highway speeds. The engines, therefore, operate inefficiently at low urban speeds with low part-load efficiency (Figure 4-8) and burn fuel while idling in traffic. Table 4-4 lists the time and fuel use shares during

[2] For hybrid systems, which are discussed next in this chapter, only the battery's specific power is critical, because discharged batteries can be recharged during operation of the internal combustion engine. High-power hybrid electric vehicle NiMH batteries currently have a specific power of about 400 W/kg and a specific energy of about 40 Wh/kg (at a 3-hour rate.) It is assumed that battery performance will improve over the next decade or so, especially in specific power, and that the goals of 800 W/kg and 50 Wh/kg are well within reach (F. R. Kalhammer, Electric Power Research Institute, personal communication, 2000). Again, lithium-ion battery technology may well surpass this goal.

vehicle stops and braking decelerations for a typical U.S. car for urban driving cycles in the United States, Japan, and Europe. The table reveals that a vehicle spends a significant amount of time and fuel during both stops and braking decelerations and that the fuel saving potential for engine idle-off is very large. If engine start-stop is designed to recover energy losses during vehicle stops only, about 12–19 percent of fuel could be saved for a vehicle operating in these cycles. The fuel savings would be even higher if engine start-stop were designed to recover all engine idling losses. This savings can be facilitated through hybridization.

Many different types of hybrid vehicles have been developed (Rovera and Mesaiti, 1999). In general, a hybrid is designed with an engine that is smaller than that needed for a similar nonhybrid car. The hybrid's smaller combustion engine (spark ignition or diesel) operates closer to its peak efficiency, which occurs near its maximum power output (a larger engine would be operating at lower efficiency at the same power output). The engine can be shut off during vehicle stops, braking decelerations, and even low-power driving, depending on the specific design of the hybrid system. A rechargeable battery system is used to provide extra power on demand. At low speeds or while the vehicle is stopped, unless the engine is shut off, the combustion engine uses excess power to recharge the battery system. The most efficient hybrid vehicle configuration also captures the regenerative energy from braking the car and uses it to recharge the battery. This type of hybrid offers major improvements in efficiency in urban driving cycles and results in lower total emissions because of the smaller engine—which is switched off at idle if the battery system is fully charged.

The disadvantages of hybrids are the extra materials and weight asso-

TABLE 4-4 Percentage of Time Spent and Fuel Consumed by a Typical U.S. Car during Vehicle Stops and Braking in Different Urban Driving Cycles

	Vehicle Stops		Vehicle Braking		Total Engine Idle	
Driving Cycle	Time	Fuel	Time	Fuel	Time	Fuel
FTP	19.2	11.6	23.6	14.2	42.8	25.8
Japan	29.2	19.1	23.1	15.2	52.3	34.3
Europe	24.9	14.5	15.7	9.2	40.6	23.7

NOTE: FTP = Federal Test Procedures (U.S.).

SOURCE: Authors' estimate; also see An et al. (2002).

ciated with a parallel electric drive system and with the battery system. Costs and some vehicle efficiency penalties are associated with the added components, and the larger the battery system, the higher the added costs and the greater the weight. For urban driving, a relatively small battery system matches the stop-and-go driving that allows frequent draw-down and recharging of the battery. In highway driving, the battery may not recharge at all, and there is the danger that it may be discharged, at which point the car loses power because it is operating on an underpowered combustion engine. For this reason, the Toyota Prius sold in the United States has both a larger engine and a larger battery system than the Japanese model.

Today's hybrid electric vehicles (HEVs) have three basic configurations:

1. The *series* HEV configuration, in which the engine drives a generator that produces electricity, which in turn powers a motor to drive the wheels and, during periods of low power demand, charges the battery. Braking energy also can be used to charge the battery. This configuration is called a series hybrid because the power flows along a single path.
2. The *parallel* HEV configuration, in which both the engine and motor drive the vehicle wheels. Because the power flows along two paths, this configuration is called a parallel system. This system also allows the engine to charge the battery on board and recover braking energy.
3. The *power-split* HEV configuration, which is closer to the parallel configuration. It differs in that a planetary gear system combined with a starter-generator can transfer power between the internal combustion engine and electric motor, both of which are coupled to the drive shaft. In this configuration, the internal combustion engine provides the primary power, with a power-split device (planetary gear with starter-generator) sending power to both the drive shaft and the electric motor. This system is sometimes called an electrically variable transmission system (Toyota, 1997).

A series HEV is technically the simplest; however, it usually requires large electrical components, and thus it is heavier and more expensive. Most hybrid city buses and heavy-duty urban trucks use series configurations. The most popular choice for today's commercial and prototype light-duty HEVs is the parallel configuration. It requires more complicated system integration, but it is lighter, more efficient, and less costly than the series system (yet more expensive than the nonhybrid technologies). The power-split system is technically more complicated, but potentially can achieve the highest efficiency.

Categories of Hybrid Vehicles

In principle, hybrid propulsion can take many forms, from slight degrees of hybridization (e.g., using an integrated starter-generator with engine start-stop capability) to designs that drive the wheels only electrically. HEVs can be classified according to the portion of their maximum propulsion power provided by an electric drive:

- *Minimal (or mini) hybrid*—fraction of onboard electric power less than about 10 percent. It can provide engine idle-off capability, but no regenerative braking and electric-only driving. The minimal hybrid usually uses a low-voltage integrated starter-generator, and its fuel economy benefit under U.S. driving conditions is about 10 percent.[3]
- *Mild and medium hybrid*—fraction of electric power ranges from 10 to 25 percent; idle-off and some regenerative braking, but no significant electric-only driving (example: Honda Insight). The fuel economy benefit of such HEVs under U.S. driving conditions is about 10–30 percent.
- *Full hybrid*—fraction of electric power ranges from 25 to 50 percent; some electric-only driving but no real trip range, and battery not designed for plug-in recharging (example: Toyota Prius). Full HEVs are sometimes called "power" hybrids. The fuel economy benefit of such HEVs under U.S. driving conditions is about 30–50 percent.

Energy HEVs (also called "charge depletion hybrids") have a useful all-electric driving range (50 miles or more) and plug-in recharging ability. But they require a large battery pack and are considerably more expensive than the other hybrids. No energy hybrids have been announced for mass production. Another reason may be that battery technologies remain too limited to provide adequate combinations of efficiency and performance even when supplemented by an engine-powered generator. Batteries are certainly a major cost factor for all hybrids, and so like pure electric vehicles energy hybrids will carry a very substantial cost premium for the foreseeable future.

A full hybrid is considered a more radical change than the other hybrids from the conventional internal combustion engine vehicle, whereas a mini or mild hybrid is considered a more natural evolution from a conventional vehicle, resulting from a historical trend of increasing vehicle onboard electric power. The vehicle electric power growth rate was about

---

[3] Hybrids can have different types of propulsion systems (spark ignition, diesel, or fuel cell) with varying efficiencies; the fuel economy benefit noted results only from the hybridization.

6 percent from 1920 to 1940, 2 percent from 1940 to 1970, and again 6 percent from 1970 to 1990 (An et al., 1999; Moore, 1999). Industry projections indicate this electrification trend will continue and probably grow. Typical U.S. cars have an onboard electric power requirement of about 1.2 kW. This electric power requirement will increase to 3–5 kW over the next few years because of the addition of features such as heated seats and windows, multimedia, water/oil pumps, power steering, HVAC (heating, ventilating, air-conditioning) fans, electromagnetic valves, and heated catalysts.

Currently, most major auto manufacturers and suppliers are working on the so-called "integrated starter-generator" system, which will increase onboard electric power to about 10–15 kW and support features such as fast crank, torque smoothing, engine idle-off, and launch assist, and a certain degree of regenerative braking. The Honda Insight hybrid vehicle falls into this category. When electric power increases to 20 kW, the vehicle's internal combustion engine can be further downsized, with such added features as electric HVAC, power assist, fast heating, and limp-home capability (if out of fuel for the engine). DaimlerChrysler's ECX2 is in this category. Vehicles with onboard electric power capability of 10–20 kW are often called mild hybrid vehicles. All major U.S. and European manufacturers are pushing for this concept of hybrid vehicle, along with standardization of a 42 V electrical system. Toyota recently released a mild hybrid option for its luxury Crown model.

When onboard electric power increases beyond 20 kW, as in the Toyota Prius and in proposed fuel cell hybrid vehicles, it finally reaches the so-called full hybrid vehicle territory. A full hybrid vehicle with a significantly downsized engine and large electric motor, combined with electrically variable transmission technologies like those developed by Toyota and Nissan (Toyota, 1997), will achieve the maximum benefit from vehicle hybridization. But consumers may find full hybrid technology too costly, placing the technology at high risk of weak customer acceptance without government or manufacturer subsidies.

Comparing Commercial and Concept HEVs

In recent years vehicle manufacturers have made great progress in developing and demonstrating commercially available and concept hybrid electric vehicles. These vehicles include commercially available gasoline hybrid cars (Toyota Prius and Honda Insight) and the diesel hybrid concept cars (Ford Prodigy, GM Precept, and DaimlerChrysler ESX3) that emerged from the Partnership for a New Generation of Vehicles (see Chapter 8). Table 4-5 summarizes some basic characteristics of selected commercial and concept HEVs in the United States (An et al., 2001b). Data sources are specified in the table because some reported figures are not always consistent among different sources, and some figures may not represent official or certified figures. For example, only Prius and Insight have fuel economy ratings certified by the U.S. Environmental Protection Agency. Other fuel economy figures are based on manufacturers' claims.

In Figure 4-12, the fraction of electric power of selected HEVs is estimated by dividing peak motor power by total combined peak motor and internal combustion engine power. Note that this figure gives only an approximate measure because the power split during operation is variable.

Assessing the Benefits of Hybrid Vehicle Fuel Economy

Because PNGV and other hybrid vehicles have such expensive drive trains, they have employed unprecedented levels of conventional vehicle fuel economy technologies such as aggressive load reduction measures. These measures include reducing weight and improving air/tire resistance as well as using lightweight materials and diesel engines to reduce weight. The choice of diesel engines for PNGV vehicles is a result of setting high efficiency goals within a five-year time horizon. The

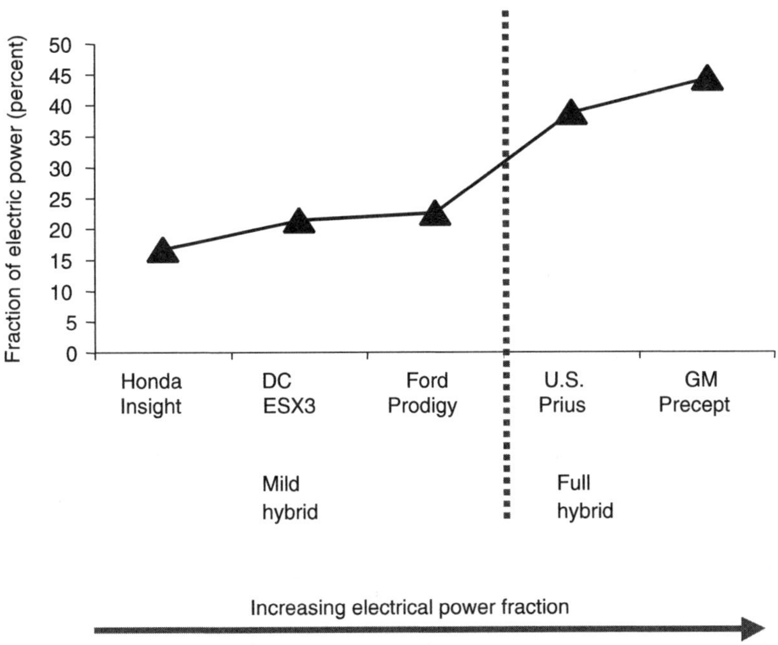

FIGURE 4-12 Fraction of electric power of selected hybrid electric vehicles. SOURCE: An et al. (2001b).

TABLE 4-5 Characteristics of Commercial and Prototype Hybrid Electric Vehicles

HEV	Type	Status	Curb Weight (lbs)	Power Plant Type	Engine Size (liters)	Engine Power (hp)	Battery Type	Motor Peak (kW)	Trans-mission Type	CAFE (mpg[a])	0–60 Time (s)
U.S. Prius	Gasoline hybrid	Commercial	2,765	SI I-4	1.5	70	NiMH	33	CVT	58	12.1
Honda Insight	Gasoline hybrid	Commercial	1,856	SI I-3	1.0	67	NiMH	10	M5	76	10.6
Ford Prodigy[b]	Diesel hybrid	Concept car	2,387	CIDI I-4	1.2	74	NiMH	16	A5	70	12.0
DC ESX3	Diesel hybrid	Concept car	2,250	CIDI I-3	1.5	74	Li-ion	15	EMAT-6	72	11.0
GM Precept[c]	Diesel hybrid	Concept car	2,590	CIDI I-3	1.3	59	NiMH	35	A4	80	11.5

[a] CAFE fuel economy rating represents combined 45/55 miles per gallon (mpg) highway/city fuel economy and is based on an unadjusted figure.

[b] On the basis of the starter-generator rated 31 kilowatts (kW) for continuous, 8 kW for three minutes, and 35 kW for three seconds (s); 16 kW is assumed for a 12-second 0–60 acceleration.

[c] The front motor is 25 kW and rear motor 10 kW. Therefore, the total motor peak power is 35 kW.

NOTE: HEV = hybrid electric vehicle; CAFE = Corporate Average Fuel Economy; SI = spark ignition; CIDI = compressed ignition direct injection; NiMH = nickel metal hydride; Li-ion = lithium-ion; CVT = continuously variable transmission; M5 = manual five gear; A5 = automatic five gear; EMAT = electro-mechanical automatic transmission; hp = horsepower.

SOURCES: U.S. Prius: *EV News* (2000); all HEVs except U.S. Prius: NRC (2000); Honda Insight: *Automotive Engineering* (1999); ESX3: *Automotive Engineering* (2000); GM Precept: Precept press release.

associated emissions and cost impacts were considered secondary in this context.

It is important to understand to what extent the gains in fuel economy are actually achieved through the conventional technologies, on the one hand, and how much are obtained through "pure" hybrid technologies and systems, on the other. Four common elements that contribute to gains in fuel economy for these commercial and concept HEVs are:

1. choice of high-efficiency diesel engines for the three PNGV hybrid vehicles
2. aggressive load reduction measures that lower vehicle air and tire drag losses, as well as overall vehicle weight
3. engine downsizing to utilize a smaller, more advanced onboard combustion engine, as well as implementation of a more advanced transmission system
4. system electrification and hybridization to utilize electric power to optimize system efficiency, turning off the engine during idling, and provision of regenerative braking.

It is not easy to estimate the fuel economy benefits of "pure" hybrid technologies. First, hybrid benefits depend strongly on the driving cycle. The fuel economy benefits of hybrid electric vehicles are much higher under stop-and-go traffic conditions than under free-flow highway driving conditions. Thus the fuel economy benefits of HEVs in China's large cities should be higher than those in the United States. Second, there is no doubt that conventional technologies include all the measures in items 1 and 2 and that hybrid technology includes all the measures in item 4. It is less clear, however, how item 3 should be categorized, because hybrid technologies usually require synergies from conventional technologies to deliver the best benefits. It would then seem that the hybrid benefits are more than item 4 alone, but somewhat less than the combination of items 3 and 4 (An et al., 2001b; DeCicco et al., 2001). Figure 4-13 presents the improvement in U.S. fuel economy by each technology element for the selected HEVs.

Issues Surrounding Fuel Cell Hybridization

Uncertainties are associated with whether hybridization would benefit fuel cell vehicles (Table 4-6). In terms of system efficiency, unlike in internal combustion engines where the engine efficiency increases with the power demand, the efficiency of fuel cell stacks peaks at a low power point and decreases as power demand approaches either the maximum or

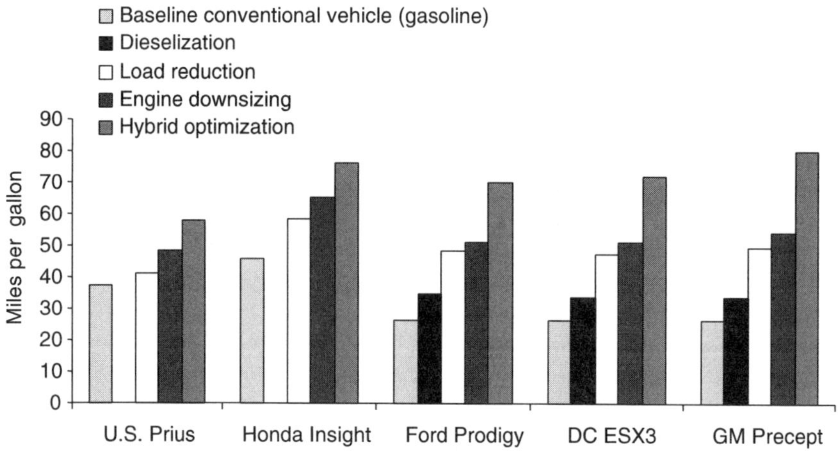

FIGURE 4-13 U.S. fuel economy (gasoline equivalent) through elements of dieselization, load reduction, engine downsizing, and hybrid optimization. SOURCE: An et al. (2001b).

minimum level. For vehicles with onboard reformers, the efficiency of the reformer's transient operation can differ dramatically from that of its steady-state operation. As for performance, the transient response time of fuel reformers is a critical issue and has a profound implication for the benefits of hybridization (the addition of batteries or other energy storage devices). In terms of system costs, the trade-off between the reduced cost associated with downsizing a fuel cell system and the added cost of batteries and other hybrid components should be assessed. The benefits and

TABLE 4-6 Trade-offs for Hybridizing Fuel Cell Vehicles

Trade-off	Fuel Cell Vehicles	Hybrid Fuel Cell/Battery
Fuel cell stack efficiency	High	Reduced because of downsizing
Regenerative braking	None	10–20+ percent in urban driving
Cold start	Slow	Rapid
Specific weight (kilograms/kilowatt)	High	High to medium
Price	Fuel cell price high, but dropping rapidly	Battery price relatively low, but not dropping as quickly as fuel cell
Overall	Long term, favorable	Near to medium term, favorable

SOURCE: Society of Automotive Engineers (2001, 2002).

trade-offs of hybridizing fuel cell vehicles not only differ from those from hybridizing conventional internal combustion engine vehicles, but also differ among different kinds of fuel cell systems.

Hybridization Summary

In spite of the many advantages of the hybrid vehicle, hybrid technologies are still quite expensive and add up to 20 percent to the cost of a vehicle. Hybrid electric vehicles are many years from taking a significant market share from the conventional internal combustion engine vehicles. In turn, the economies of scale needed to reduce their cost disadvantage probably will not be realized for many years. In the meantime, some of the first steps toward hybrid technologies—the integrated starter-generator, for example—will begin to appear in commercial products. In the United States, expensive features such as hybridization are more likely to appear in higher-priced (or subsidized) vehicles before they make their way into the wider commercial markets.

SYSTEM INTEGRATION AND MANUFACTURING

The preceding sections have described the status of the major components of a modern automobile. In China, many car components are already being manufactured for assembly in local plants and for export. Moreover, local vendors are involved in technology transfer and improvement on the component level, although entry into the World Trade Organization (WTO) will create pressure on component manufacturers to be more cost-competitive. Yet car design requires sophisticated techniques for integrating all these components—and ultimately for satisfying customer requirements for performance, safety, comfort, and convenience—to produce a cost-competitive, reliable vehicle. The leading global automakers have had years of experience in producing and marketing cars, and they continue to refine their system design and integration skills in order to maintain or improve their competitive position in the marketplace. Consequently, these skills are considered highly proprietary and are carefully guarded.

At present, China's automotive industry is well behind global automotive industry leaders in design and system integration capabilities. China has some world-class auto production plants, but these rely on the expertise of foreign partners who are reluctant to transfer this expertise to their Chinese partners. If China is to compete in the future as a world-class auto manufacturer, it must build expertise in these critical areas. In future joint ventures and similar partnerships with foreign automakers, it will be important for the Chinese partner to demand training and full inclusion in design and integration activities—at least for cars that will be developed for the Chinese market. Because Chinese automakers are likely to better understand Chinese consumers and their buying patterns, foreign partners may be willing to relax proprietary concerns in the codevelopment of models that meet the needs of Chinese consumers. Yet under WTO such partners also can compete directly in Chinese markets. Therefore, the Chinese should be prepared to bring some special capabilities into partnerships in exchange for training and knowledge transfer.

Manufacturing Implications

World-class producers of vehicles must not only design high-quality vehicles, but also produce them in efficient facilities. In any discussion of manufacturing, it is useful to separate the vehicle assembly operations from the manufacture of major components such as engines and transmissions. Major components are normally manufactured in high-capacity, highly automated plants. Because a large fraction of the tools found in these facilities are made by a few worldwide machine tool manufacturers, any facilities built in China can possess the latest technologies. Furthermore, because of the high cost of the facilities, several vehicle manufacturers may use the same components purchased from external suppliers, allowing an economy of scale in production that benefits all participants. As noted in Chapter 3, component suppliers Robert Bosch, Engelhard, and Corning already have subsidiaries in China that are following this pattern. Whether the Chinese industry chooses to compete with the large component suppliers or to allow foreign manufacturers to dominate this part of the market is largely an economic issue and not one of technical capability.

China has experience with vehicle assembly in modern plants, but these plants have been mostly ones transplanted from overseas through joint venture partners. Because the volumes for the Chinese car market will not at first be large for any single manufacturer, these facilities must be flexible enough to allow the assembly of more than one vehicle type. This flexible technology is just appearing in foreign facilities, but it is expected to find greater uses as the market continues to fracture into smaller niches. Besides, as just noted, the latest tools for an assembly plant can be purchased from the worldwide machine tool industry.

Unfortunately, even having the latest tools in the most modern facilities does not guarantee efficient operation. Japanese manufacturers, through their highly developed "lean" manufacturing processes and through the colocation of major component and assembly plants, are able to assemble a vehicle in about half the number of worker-hours characteristic of American manufacturers (Dassbach, 1994). Through years of sustained effort, Japanese manufacturers also have developed a reputation for the highest quality products. To compete worldwide, the Chinese automotive industry will need to develop the management capability to operate its factories in the most efficient manner while paying close attention to quality. China has a major asset in its intelligent, hard-working people, who provide skilled labor at wages that are low relative to the average in the member countries of the Organisation for Economic Co-operation and Development (OECD), but it will need to examine carefully its capability to manage that workforce in the most efficient manner.

Choosing Competitive Technologies

As discussed in Chapter 3, a variety of cars are already being produced successfully in China. Most of these employ foreign technology, adapted to some extent to meet special Chinese needs, through a variety of partnerships and joint ventures with Chinese companies. If the Chinese automotive industry wishes to compete in open international markets or to compete with products coming to China from these markets, it must find technologies or market niches in which it is able to compete successfully. In Figure 4-14, the estimates by Weiss et al. (2000) are replotted to show life cycle energy as a function of life cycle cost for the technologies considered. If China is to compete successfully in the manufacture of lower-cost cars, it would seem prudent to select technologies that do not start out with a potentially significant cost penalty.

As shown in Figure 4-14, the overlap in the energy use and cost projections for technology alternatives is large, especially for the advanced technologies for which the uncertainty ranges are largest. These uncertainty ranges will be reduced over the next decade or so as advanced technologies mature from the present stage of development. Therefore, much more risk is associated with future competitiveness if major investments are made now by China to develop hydrogen fuel cell or electric cars, or even full hybrids, rather than in the future. A near-term focus on more conventional technologies still allows many opportunities for innovations that will enhance the competitiveness of a China car.

APPENDIX: OTHER TYPES OF FUEL CELLS

Of the many types of fuel cell, some are being developed for stationary power generation. These systems are more appropriate for steady-state operation because they require the maintenance of a special environ-

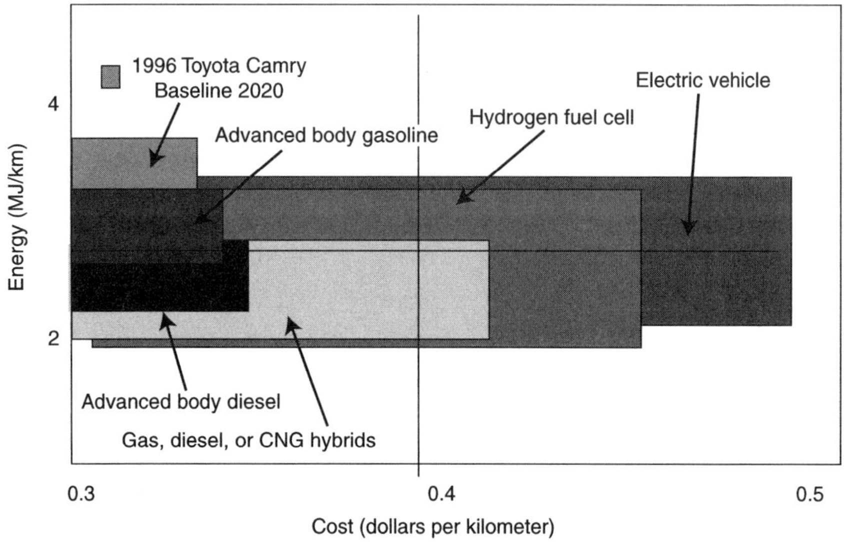

FIGURE 4-14 Estimated energy versus cost ranges for selected technologies per kilometer. NOTE: Shaded areas represented uncertainties in prediction; CNG = compressed natural gas. SOURCE: Weiss et al. (2002).

ment. Thus such systems may not yet meet the rapid start-up requirement of an automobile and are likely to perform inefficiently in a lightly loaded automotive duty cycle. Descriptions of some of the fuel cells likely to find commercial uses over the next several decades follow:

- *Phosphoric acid.* This type of fuel cell, which is commercially available, is used mostly for stationary power sources. Phosphoric acid fuel cells generate electricity at more than 40 percent efficiency—and nearly 85 percent of the steam this fuel cell produces is used for cogeneration. Operating temperatures are in the vicinity of 400°F or 200°C.
- *Molten carbonate***.** Molten carbonate fuel cells, which promise high fuel-to-electricity efficiencies, operate at about 1,200°F or 650°C. To date, molten carbonate fuel cells have been operated using hydrogen, carbon monoxide, natural gas, propane, landfill gas, marine diesel, and simulated coal gasification products. Molten carbonate fuel cells of 10 kW to 2 megawatts (MW) have been tested on a variety of fuels. Carbonate fuel cells for stationary applications have been successfully demonstrated in Japan and Italy.
- *Solid oxide.* Another highly promising fuel cell, the solid oxide fuel cell (SOFC), could be used in large, high-power applications, including

industrial and large-scale central electricity generating stations. Some researchers also foresee the use of solid oxide in motor vehicles and are developing fuel cell auxiliary power units with SOFCs. A solid oxide system usually uses a hard ceramic material instead of a liquid electrolyte, allowing operating temperatures to reach 1,800°F. Power-generating efficiencies could reach 60 percent. One type of SOFC uses an array of meter-long tubes, and other variations include a compressed disc that resembles the lid of a soup can. Tubular SOFC designs are closer to commercialization and are being produced by several companies. Demonstrations of tubular SOFC technology have produced as much as 220 kW.

Still other fuel cells are not likely to be commercialized in the next few decades:

- *Alkaline.* Long used by the U.S. National Aeronautics and Space Administration (NASA) on space missions, these fuel cells can achieve power-generating efficiencies of up to 70 percent. They use alkaline potassium hydroxide as the electrolyte. Until recently, these fuel cells were too costly for commercial applications, but several companies are examining ways to reduce costs and improve operating flexibility.
- *Direct methanol fuel cells.* These cells are similar to the proton exchange membrane fuel cell (see the section Emerging Propulsion Technologies) in that they use a polymer membrane as the electrolyte. However, in the direct methanol fuel cell, the anode catalyst itself draws the hydrogen from the liquid methanol, eliminating the need for a fuel reformer. Efficiencies of about 40 percent are expected with this type of fuel cell, which would typically operate at a temperature of between 120° and 190°F or 50° and 90°C. Higher efficiencies are achieved at higher temperatures.
- *Regenerative fuel cells.* Still a very young member of the fuel cell family, regenerative fuel cells would be attractive as a closed-loop form of power generation, in conjunction with a solar power source. Solar power can be converted directly to electricity, but there may be applications where the energy losses associated with a regenerative fuel cell are justified. During solar collection, water is separated into hydrogen and oxygen by a solar-powered electrolyzer. The hydrogen and oxygen are then available to be fed into the fuel cell, which is able to generate electricity, heat, and water when direct solar power is not available. The water is then recirculated back to the solar-powered electrolyzer and the process begins again. NASA and others are currently conducting research on these types of fuel cells.

REFERENCES

An, F., F. Stodolsky, and D. Santini. 1999. Hybrid Options for Light-Duty Vehicles. SAE Technical Paper No. 1999-01-2929. Society of Automotive Engineers.

An, F., J. DeCicco, and M. Ross. 2001a. Assessing the Fuel Economy Potential of Light-Duty Vehicles. SAE Technical Paper No. 2001- 01FTT-31, SAE Special Publication (SP-1637) on New Energy Systems and Environmental Impacts. Society of Automotive Engineers.

An, F., A. Vyas, J. Anderson, and D. Santini. 2001b. Evaluating Commercial and Prototype HEVs. SAE Technical Paper No. 2001-01-0951. Society of Automotive Engineers.

An, F., D. Freidman, and M. Ross. 2002. Near-Term Fuel Economy Potential for Light-Duty Trucks. SAE Technical Paper No. 2002-01-1900. Society of Automotive Engineers.

Ashley, S. 2001. Fuel cells start to look real. Automotive Engineering (March).

Automotive Engineering. 1996. Life cycle analysis: Getting the total picture on vehicle engineering alternatives. (March):49–52.

———. 1999. ESX3. (October):55.

———. 2000. Dodge's mild hybrid. (May):32.

Chinese State Environmental Protection Agency. 1999. Technical Policy on Prevention and Control of Motor Vehicle Pollution. Beijing.

Dassbach, C. 1994. Where is North American automobile production headed? Low-wage lean production. Electronic Journal of Sociology 1(1).

Davis, B. H. 1999. Transportation Energy Data Book: Ed. 19. ORNL-6958. Oak Ridge National Laboratory, September.

Dec, J. E. 1997. A Conceptual Model of DI Diesel Combustion Based on Laser-Sheet Imaging. SAE Technical Paper No. 970873. Society of Automotive Engineers.

DeCicco, J., F. An, and M. Ross. 2001. Technology Options for Improving the Fuel Economy of U.S. Cars and Light Trucks by 2010–2015. American Council for an Energy Efficient Economy, Washington, D.C., April.

De Witt, M. J., and C. Z. Wan. 2000. Establishing capabilities to evaluate reductants for NO_x adsorber technologies. Paper presented at the 6th Diesel Engine Emissions Reduction Workshop, August.

Ehsani, M., A. Emade, and H. Gao. 2001. 42V Automotive Power Systems. SAE Technical Paper No. 2001-01-2465. Presented at the Future Transportation Technology Conference, Costa Mesa, Calif., August.

EV News. 2000. June.

Flynn, P. F. 2000. Diesels—Promise and issues. Paper presented at the 6th Diesel Engine Emissions Reduction Workshop, August.

———. 2001. Diesels 2007—Promise and problems. Paper presented at the 7th Diesel Engine Emissions Reduction Workshop, August.

Flynn, P. F., R. P. Durrett, G. L. Hunter, L. A. Farrell, and O. C. Akinyemi. 2000. Minimum Engine Flame Temperature Impacts on Diesel and Spark-Ignition NO_x Production. SAE Technical Paper No. 2000-01-1177. Society of Automotive Engineers.

Flynn, P. F., R. P. Durrett, G. L. Hunter, A. O. zur Loye, O. C. Akinyemi, J. E. Dec, and C. K. Westbrook. 1999. Diesel Combustion: An Integrated View Combining Laser Diagnostics, Chemical Kinetics, and Empirical Validation. SAE Technical Paper No. 1999-01-0509. Society of Automotive Engineers.

General Motors Corporation (GMC). 2001. Well-to-Wheels Energy Use and Greenhouse Gas Emissions of Advanced Fuel/Vehicle Systems: North American Analysis. Executive Summary Report, Detroit, April.

Höhlein, B., J. Nitsch, and U. Wagner. 1998. Brennstoffzellen-Studie (Fuel Cell Studies): Vorhaben Nr. 686, Ganzheitliche Systemuntersuchung zur Energiewandlung durch Brennstoffzellen. Forschungsvereinigung Verbrennungskraftmaschinen e.V., Frankfurt am Main.

Horton, E. J., and W. D. Compton. 1984. Technological trends in automobiles. Science 225:588.

Jost, K. 2002. Spark-ignition engine trends. Automotive Engineering International (January).

Kummer, J. T. 1981. Catalysts for automobile engine control. Progress in Energy and Combustion Science 6:177–199.

Moore, B. 1999. Advanced energy systems. Paper presented at the SAE Topical Technology Workshop on Hybrid Electric Vehicles: Here and Now (TOPTEC). SAE (Society of Automotive Engineers) International, Warrendale, Pa.

Naber, J. D., and D. L. Siebers. 1996. Effects of Gas Density and Vaporization on Penetration and Dispersion of Diesel Sprays. SAE Technical Paper No. 960034. Society of Automotive Engineers.

National Research Council (NRC). 2000. Review of the Research Program of the Partnership for a New Generation of Vehicles. Sixth Report. Washington, D.C.: National Academy Press.

———. 2002. Effectiveness and Impact of Corporate Average Fuel Economy (CAFE) Standards. Washington, D.C.: National Academies Press.

Office of Technology Assessment. 1995. Advanced Automotive Technologies: Visions of a Super-Efficient Family Car. OTA-ETI-638. Washington, D.C.: Government Printing Office.

Ogden, J. M., M. M. Steinbugler, and T. G. Kreutz. 1999. A comparison of hydrogen, methanol, and gasoline as fuels for fuel cell vehicles: Implications for vehicle design and infrastructure development. Journal of Power Sources 79:143–168.

Pembina Institute. 2000. Climate-Friendly Hydrogen Fuel: A Comparison of the Life-Cycle Greenhouse Gas Emissions for Selected Fuel Cell Vehicle Hydrogen Production Systems. Pembina Institute, Drayton Valley, Alberta, March.

Rovera, G., and D. Mesaiti. 1999. The general purpose hybrid vehicles. Paper presented at the SAE Topical Technology Workshop on Hybrid Electric Vehicles: Here and Now (TOPTEC). SAE (Society of Automotive Engineers) International, Warrendale, Pa.

Sher, E., ed. 1998. Handbook of Air Pollution from Internal Combustion Engines: Pollutant Formation and Control. Boston: Academic Press.

Siebers, D. L. 1999. Scaling liquid-phase fuel penetration in diesel sprays based on mixing limited vaporization. Paper presented at 1999 Congress of the Society of Automotive Engineers, Detroit, February.

Sierra Research. 1997. Automotive fuel economy potential using cost-effective design changes. Sierra Research, Buffalo, N.Y.

Singh, M., R. Cuenca, J. Formento, L. Gaines, B. Marr, D. Santini, M. Wang, S. Adelman, D. Kline, J. Mark, J. Ohi, N. Rau, S. Freeman, K. Humphreys, and M. Placet. 1998. Total Energy Cycle Assessment of Electric and Conventional Vehicles: An Energy and Environmental Analysis. ANL/ES/RP-96387, Argonne National Laboratory.

Society of Automotive Engineers. 2001. Fuel Cell Power for Transportation 2001. SAE SP-1589.

———. 2002. Fuel Cell Power for Transportation 2002. SAE SP-1691.

Stodolsky, F., L. Gaines, C. Marshall, and F. An. 1999. Total Fuel Cycle Impacts of Advanced Vehicles. SAE Technical Paper No. 1999-01-0322. Society of Automotive Engineers.

Toyota. 1997. Prius Product Information 2000. Toyota Hybrid System, Toyota Press Information 1997.

U.S. Council for Automotive Research. 2000. U.S. Advanced Battery Consortium Battery Criteria: Commercialization. Online. USCAR. Available at www.USCAR.org. Accessed May 2000.

U.S. Department of Energy. 1999. Hybrid electric vehicle component information. Online. Available at www.hev.doe.gov/components/batteries.html. Accessed June 13, 2002.

———. 2001. Fuel Cells for Transportation. 2001 Annual Progress Report, December.

Wang, M. Q. 1999. GREET 1.5—Transportation Fuel-Cycle Model, Volume I: Methodology, Development, Use, and Results. ANL/ESD-39. Argonne National Laboratory, August.

Wang, M. Q., and H.-S. Huang. 1999. A Full Fuel-Cycle Analysis of Energy and Emissions Impacts of Transportation Fuels Produced from Natural Gas. ANL/ESD-40. Argonne National Laboratory, December.

Weiss, M. A., J. B. Heywood, E. M. Drake, A. Schafer, and F. F. AuYeung. 2000. On the Road in 2020: A Life-Cycle Analysis of New Automobile Technologies. MIT Energy Laboratory Report #MIT-EL-00-003. Massachusetts Institute of Technology, Cambridge.

Wengel, J., and E. Schirrmeister, eds. 2000. The Innovation Process from the Internal Combustion Engine to Fuel Cells. Fraunhofer-Institute, Karlsruhe, Germany, February.

Westbrook, C. K., and L. K. Chase. 1988. Chemical Kinetics and Thermochemical Data for Combustion Applications. Lawrence Livermore National Laboratories Report UCID-17833.

Westbrook, C. K., W. J. Pitz, and W. R. Leppard. 1991.The Autoignition Chemistry of Paraffinic Fuels and Pro-Knock and Anti-Knock Additives: A Detailed Chemical Kinetic Study. SAE Technical Paper No. 912314. Society of Automotive Engineers.

Williams, R. 1998. Fuel cells, coal, and China. Paper presented at the Ninth Annual U.S. Hydrogen Meeting: Implementing a Global Solution, Washington, D.C.

Zhan R., R. Hurley, W. Han, J. A. Lymburner, D. Schuetzle, J. Li, G. Wu, L. Xu, H. S. Gandhi, R. Huang, and D. Deng. 2001. The Development of Low Precious-Metal, Rare-Earth Oxide (Reo) Catalysts for Vehicle Emission Control in Emerging Markets. SAE Technical Paper No. 2001-01-0225. Presented at the SAE 2001 World Congress, March, Detroit. Society of Automotive Engineers.

5

Energy and Fuels

Energy is playing a key role in the rapid development of China. Industrialization and growth of the country's gross domestic product (GDP) depend heavily on the availability of affordable and reliable energy. The transportation sector is dependent on such energy as well. As incomes rise people generally seem to travel farther (Schafer, 1998). With the recent rise in per capita income in China, more people are able to afford cars and want the personal benefits that automobile ownership provides. As the automobile fleet grows, the demand for the fuels it needs and the supporting supply and distribution infrastructure for those fuels also will increase (see Chapters 3 and 4).

As shown in Table 5-1, today the Chinese use much less energy per capita than citizens of the member countries of the Organisation for Economic Co-operation and Development, or OECD (U.S. DOE, 1999). The average Chinese citizen, at 0.6 ton of oil equivalent (TOE)[1] per capita per year, uses about 8 percent of the energy consumed by the average U.S. citizen and about 15 percent of the energy used by the average citizen of Japan or Germany. The high U.S. energy consumption is linked in part to the greater use of energy in the United States for transportation, which is linked in turn to lower population density. Globally, an average of 1.4 TOE per capita per year is consumed. The challenge, then, is how to pro-

[1] One ton of oil equivalent equals 40 million Btus (MBtu), 7.35 barrels (bbl) of oil equivalent, 0.96 kiloliters (kl) of oil equivalent, 11.8 megawatt hours (MWh), and 4.26×10^{10} joules (J).

TABLE 5-1 Average per Capita Energy Use for Selected Countries, 1999 (tons of oil equivalent [TOE] per person per year)

Country	TOE/Person/Year	Country	TOE/Person/Year
United States	8.0	Mexico	1.2
Canada	7.3	Brazil	0.7
Norway	5.5	China	0.6
Russia	4.2	India	0.3
Japan	4.0	Africa	0.14
Germany	4.0	Bangladesh	0.08
World average = 1.4 TOE/person/year			

SOURCE: U.S. DOE (1999).

vide a source of inexpensive energy to the developing countries as they seek to become more developed and yet remain mindful of concerns about excessive dependency on oil imports and the need to limit global greenhouse gas (GHG) emissions.

From a strategic point of view, a shortage of domestic oil is a barrier to the development of an automotive industry. Motor vehicles in China consume 85 percent of the country's gasoline output and 42 percent of its diesel output. In 1995 China's demand for oil was 3.0 million barrels per day (mbd) or 147 million metric tons (MMT) per year, growing to 4.5 mbd (220 MMT) in 2000, and projected to reach 5.2 mbd (250 MMT) by 2005 (Chen, 2001). In 2000 imports of petroleum were 70 MMT, and an annual increase in imports of at least 10 MMT per year is anticipated in the short term. According to predictions, by 2010 China will need 270–310 MMT of crude oil per year (Yang et al., 1997). Unfortunately, the domestic supply will reach just 165–200 MMT per year, and the deficit of 105–110 MMT must be imported.

The rapid growth in the vehicle sector is the primary force driving China's rapid shift from being a net petroleum-exporting country to a net importer. This shift not only creates concerns about China's energy security and balance of payments, but also increasingly strains China's refinery sector, which traditionally has been largely able to provide the country's own refined product needs, using a refining network set up for indigenous heavy, sweet crudes (Histon, 2001). A particular concern is the high sulfur content of imported crude oil compared with that of domestic crude. Because the Chinese refineries were built to process the relatively low-sulfur domestic crude, the available hydrodesulfurization capacity is limited.

The quality of fuels is inextricably linked to the regulations for vehicle emissions performance. China has decided to follow the pollution control

TABLE 5-2 European Union Fuel Specification Limits

Petrol/Gasoline	2000	2005	Diesel	2000	2005
RVP summer kPa, max.	60	—	Cetane number, min.	51	—
Aromatics, % by vol. max.	42	35	Density 15°C kg/m^3, max.	845	—
Benzene, % by vol. max.	1	—	Distillation 95% by vol. °C, max.	360	—
Olefins, % by vol. max.	18	—	Polyaromatics, % by vol. max.	11	—
Oxygen, % by mass max.	2.7	—	Sulfur, ppm max.	350	50
Sulfur, ppm	150	50			

NOTE: Dashes signify that no changes to existing levels have yet been issued for implementation in 2005. RVP = Reid Vapor Pressure; kPa = kilopascals (1 atmosphere of pressure equals about 100 kPa); kg/m^3 = kilograms per cubic meter; ppm = parts per million.

SOURCE: Directive 98/70/EC of the European Parliament and of the Council of 13 October 1998, amending Council Directive 93/12/EEC.

strategies of the European Union (EU), and so, as noted in Chapters 4 and 7, will upgrade its fuel quality, including further reductions in sulfur, to meet those emissions standards. Table 5-2 shows the fuel specification standards that are now in effect in the EU and that will be required for fuels sold in 2005. After an extensive consultation process, the European Union Commission initially proposed requiring the introduction of gasoline and diesel fuels with less sulfur than 10 parts per million (ppm) or 0.001 percent by mass as early as 2005, with a complete shift to these low-sulfur fuels by 2011.[2] Because lower sulfur levels in gasoline and diesel fuel are preconditions for the introduction of advanced vehicle technologies that are able to comply with future European Emission Standard III

[2] In November 2001 the European Parliament, in its first reading, called for a complete conversion to fuels with a maximum sulfur content of 10 ppm (0.001 percent by mass) by 2008, and in December the Council of Ministers decided on a "common position" of complete conversion by 2009. The issue has now gone back to the Parliament for its second reading.

(Euro III) and Euro IV standards, China will have to substantially upgrade its refineries. (A more extensive discussion of the environmental implications of fuel quality is presented in Chapter 7.)

In addition, the fuel efficiency of most Chinese cars today is poorer than that of cars of comparable weight and size in the industrialized countries. Unless fuel economy is improved in the future, even greater strains will be placed on the refinery sector. As noted in Chapter 3, China is anticipating at a minimum a threefold increase in its vehicle fleet, not including motorcycles, between 2002 and 2020. The automobile fleet in particular is expected to increase by a factor of between four and five within the same time period. Based on the vehicle characteristics in the tenth five-year plan, it is estimated that total fuel consumption will more than double by 2020 despite a gradual improvement in gasoline vehicle fuel efficiency and an increase in the use of more efficient diesel technology. Table 5-3 reveals the effects of both moderate and more aggressive attention to vehicle fuel economy over the period 2000–2015. Year 2000 is used as the base case. Case 1 assumes that starting in 2005 the fuel economy of all new gasoline-fueled cars and light trucks improves by 2 percent a year. Building on this, case 2 assumes that starting in model year 2010, a small fraction of highly efficient cars and light trucks, increasing by 5 percent a year, achieves fuel consumption of 80 miles per gallon (mpg). As illustrated, these fuel economy standards would start to reduce the growth in fuel consumption but would need several more years to have their full impact. Although the vehicles in the Chinese fleet today may be somewhat less efficient than those in the same size range in the U.S. fleet, the real inefficiency lies in the fuel wasted because of urban congestion. Such

TABLE 5-3 Influences of Vehicle Efficiency Improvements on Future Fuel Consumption in China

Assumptions	2000	2005	2010	2015	2020
Base case: Year 2000 trends continue	1.00	1.48	2.18	3.29	4.97
Case 1: Fuel economy of fleet improves at rate of 2% per year	1.00	1.47	2.14	3.13	4.49
Case 2: Same as above case, but with further addition of 80 mpg cars to the fleet at a rate of 5% of new cars per year	1.00	1.47	2.14	2.98	3.97

NOTE: Table shows ratio of fuel consumed in future to fuel consumed in 2000 for China's light-duty vehicle fleet. Light-duty vehicles include light-duty trucks, light vans, and a variety of jeeps, lighter than 3.5 tons. SOURCE: Calculations by Michael P. Walsh.

congestion may increase if the urban road infrastructure does not keep pace with the growth in the number of vehicles.

The overall implication of this analysis is that improvements in vehicle efficiency will help reduce fuel demand as the Chinese car fleet expands over the coming decades, but even these improvements will not offset the increased use of petroleum. Smaller vehicles with lower average fuel consumption may reduce fuel consumption, but to succeed such vehicles will have to be attractive to Chinese customers.

CHINA'S FUEL INDUSTRY

Chinese crude oils have less naphtha, the feedstock for the catalytic reforming process, than most foreign crudes. Therefore, one feature of the Chinese petroleum refining industry is that its catalytic cracking capacity is much greater than its capacities for catalytic reforming and catalytic hydrotreating. The principal characteristics of Chinese gasoline are a high olefin and sulfur content; likewise, Chinese diesel fuel has a high sulfur content. Because environmental protection regulations have become stricter in recent years in China, the quality of Chinese fuels has improved markedly; the olefin content of gasoline and the sulfur content of diesel fuel have fallen. New, higher-quality gasoline and diesel fuel specifications are forthcoming.

More broadly, China's petroleum industry is becoming more open to foreign participation, and China's accession to the World Trade Organization (WTO) will spur further changes. The China National Petroleum Corporation (CNPC), which produces about 64 percent of China's crude, had signed more than $1 billion (RMB8.3 billion) in contracts with foreign countries by 1997, and these and more recent partnerships are accelerating domestic exploration and resulting in significant additions to reserves. However, because importation will require further contractual obligations, the CNPC has signed contracts of crude exploitation with countries such as Iraq, Kazakhstan, Sudan, Peru, and Venezuela.

Petroleum Refining Industry

CNPC and the China Petroleum and Chemical Corporation (SINOPEC) were established after a reorganization of the Chinese petroleum industry in 1998. Subsidiaries of both companies undertake crude oil exploration and production, as well as refining. China has a petroleum processing capacity of 250 MMT annually, which places it fourth in the world, after the United States, the former Soviet Union, and Japan. In 2000 China processed 202 MMT of crude oils, producing 41 MMT of gasoline and 70 MMT of diesel fuel, which met the current domestic demand.

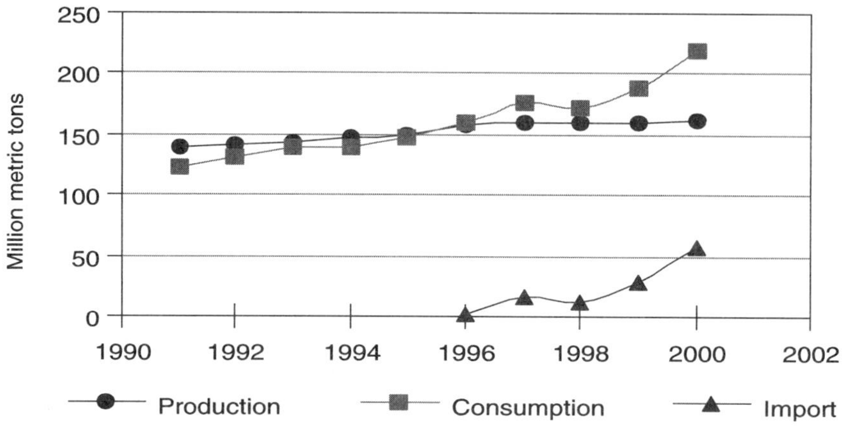

FIGURE 5-1 China's annual crude oil production, consumption, and imports, 1991–2000. SOURCE: Chen (2001).

SINOPEC's crude oil runs account for 52 percent of the national total. Figure 5-1 shows the production, consumption, and importation rates for crude oil from 1991 to 2000. Figure 5-2 indicates that China's production of diesel oil, and its growth rate, is higher than that for gasoline. From 1991 to 2000 the consumption of diesel fuel increased rapidly because of the higher numbers of light- and medium-duty diesel trucks and diesel agricultural vehicles and equipment.

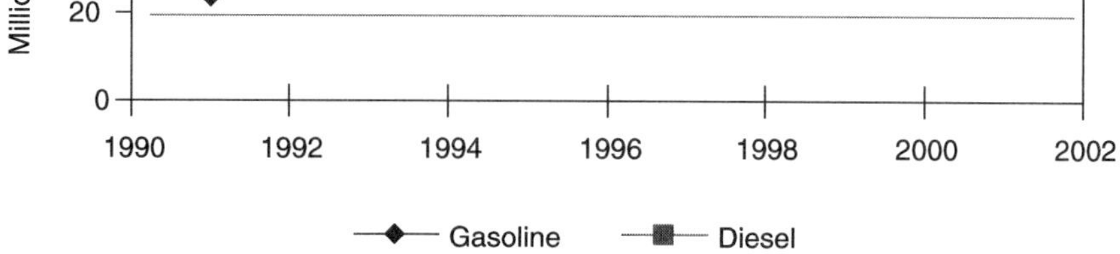

FIGURE 5-2 China's production of gasoline and diesel fuel, 1991–2000. SOURCE: Ruan (2001).

TABLE 5-4 China's Refining Capacities

Refining Process	Capacity (thousand metric tons per year)	Percent
Distillation	249,450	100.0
Catalytic reforming	14,877	6.0
Catalytic cracking	87,655	35.1
Catalytic hydrocracking	12,920	5.2
Hydrorefining of residue	5,200	2.1
Hydrotreating	35,105	14.1

NOTE: The percentages shown are based on distillation capacity.

SOURCE: Data collected by Qiu Yansheng, Research Institute of Petroleum Processing.

China's high catalytic cracking capacity constitutes 35 percent of its total distillation capacity (see Table 5-4). Table 5-4 also reveals that China has a smaller catalytic reforming capacity (6 percent) than that of the United States (18.7 percent) because of the lower levels of naphtha in its domestic crude. Furthermore, China's catalytic hydrocracking and catalytic hydrotreating capacities are not sufficient. As a result, at the moment China's refining industry has a limited ability to supply advanced gasoline and diesel fuel components. As noted earlier, this industry is geared to handle the quality of Chinese crude, most of which is heavy, low-sulfur, and waxy, and therefore refiners lack the capacity to process high-sulfur crude.

At present, China has about 50 refineries with crude oil input capacities of greater than 1 MMT per year and an average capacity of 4.76 MMT per year. About 70 additional refineries have a capacity of less than 1 MMT per year.

Growing concerns in China about the environmental impacts of rising oil consumption have led to investments in new refining technologies and the revision of product specifications. Among the earliest policy targets was eliminating the 66 and 70 MON (Motor Octane Number) specifications for gasoline, raising the new minimum to 90 RON (Research Octane Number), and eliminating alkyl-lead additives for octane enhancement through the increased use of alkylates, reformate, and methyl tertiary-butyl ether (MTBE) and other oxygenates in gasoline blending. New unleaded specifications for 93 and 95 octane (RON) gasoline were added as well. Methyl cyclopentadienyl manganese tricarbonyl (MMT) is used as an octane enhancer by about 50 percent of China's refineries, but it is probably not a good solution because of its cost, heavy

TABLE 5-5 Consumption Patterns for Diesel Fuels in China, 1995–2000 (percent by mass)

Consumer	1995	1997	2000
Agriculture	26.2	23.4	21.8
Fisheries	11.9	11.4	10.9
Transportation	48.1	50.0	57.1
Highways	29.8	34.4	41.8
Railways	10.3	8.8	8.5
Marine	8.0	6.8	6.8
Electricity	6.9	9.5	5.1
Others	6.9	5.7	5.1
Total	100	100	100

SOURCE: Data collected by Qiu Yansheng, Research Institute of Petroleum Processing.

metal content, and adverse effect on hydrocarbon emissions from catalyst-controlled cars (Owen and Coley, 1995). The Chinese industry is using MMT as a transitional measure and expects to eliminate its use in the future.

In China, almost all gasoline is used in road transportation, but only half of diesel fuels (Table 5-5). A survey of Chinese gasolines in 1999 revealed their high olefin and sulfur content. Olefins in gasoline cause deposits in the intake system and fuel injectors of gasoline engines and an increase in the photochemical reaction activity of engine exhaust gas. Sulfur compounds poison the catalysts of the exhaust gas emissions control systems. Recently, several measures have been adopted in Chinese refineries to improve gasoline quality, including adding innovative catalytic cracking technology, increasing catalytic reforming capacity, and installing hydrodesulfurization facilities.

Gasoline Specifications

Gasoline specifications are driven by emissions standards. Gasoline additives aimed at improving performance have a somewhat checkered history. Lead has been used since the 1920s as an octane enhancer, but it was later found to cause severe health problems and to poison catalysts. It has since been phased out, although some is still used in developing coun-

TABLE 5-6 Unleaded Petrol Specification for Motor Vehicles, July 2000 (GB 17930-1999)

Item	Limit		
Research Octane Number min.	90	93	95
Antiknock index min.	85	88	90
Lead, g/liter max.	0.005		
Sulfur, ppm max.	1,000		
Benzene, % by vol. max.	2.5		
Aromatics, % by vol. max.	40		
Olefins, % by vol. max.	35		

NOTE: ppm = parts per million.

SOURCE: China State Bureau of Quality and Technical Supervision (1999).

tries. More recently, MTBE was a required component of reformulated gasoline in the United States, but it was shown to contaminate groundwater. The debate continues over whether oxygenate additives to gasoline significantly improve emissions performance, and ethanol is being evaluated as a possible substitute in China.

The costs of improved fuels to the consumer (before taxes) will not be materially different from the present ones, adjusted for inflation, because the oil industry has been creative in finding technologies to produce cleaner fuels at little or no additional cost. However, new large capital investments may be needed for the Chinese oil industry. Some special grade of higher-purity gasoline designed for fuel cell vehicles fitted with gasoline reformers may appear, but it need not be radically different from the cleaner gasoline that will evolve for use in cars with internal combustion engines. Fuel prices at the pump, on the other hand, may be influenced in the future by tax policies, market manipulations, or other exogenous factors, so they may be more variable. Fuel production costs, however, are an important factor in examining the competitive introduction of alternative fuels.

In recent years, environmental protection has received considerable attention in China, especially in the mega cities of Beijing, Shanghai, and Guangzhou. The resulting more stringent regulations for engine emissions will have a strong impact on the fuel industry. The significant difference between the new specifications of July 2000 and the original ones is the regulatory limitation of olefin and aromatic contents. Although the maximum allowable limits of olefin and sulfur content are still higher than those for U.S. gasoline, the standards represent major progress given the current status of the Chinese petroleum refining industry. Table 5-6 gives

China's newest gasoline specification, GB 17930-1999 (China State Bureau of Quality and Technical Supervision, 1999). And Chinese gasoline quality will continue to improve; the next goal is to reduce sulfur content to 200 ppm. The total amount of olefins and aromatics in the gasoline pool are limited to 60 percent by volume maximum, and the limit of olefin content is 35 percent by volume maximum. These changes are included in SINOPEC's gasoline specification for city automobiles, Q/SHR 007-2000. New gasoline specifications will be implemented in 2003.

It is noteworthy that SINOPEC has a reference specification for exported unleaded gasoline that currently meets or exceeds the new domestic gasoline specifications. As the Chinese refining industry moves toward world standards, the complications introduced by having to produce different fuels to meet different quality specifications (other than those changes needed to maintain good performance under regional and seasonal climate variations) should gradually disappear.

Diesel Fuel

Chinese diesel fuels contain a small portion of hydrotreated components, which results in high sulfur content. Table 5-7 reveals that the sulfur content of Chinese diesel fuels increased between 1995 and 1997 because of an increase in imported sour crude oils. The sulfur content then decreased after 1997, when catalytic hydrotreating operations were improved. Although there is general agreement that lower sulfur will reduce particulate emissions, it is not yet clear whether changes in other properties can substantially reduce harmful emissions from current heavy-duty diesel engines. However, revised fuel specifications would be important in new diesel engine designs using exhaust gas aftertreatment, or in light-duty diesel engines for passenger cars. Modifications to reduce emissions to very low levels may reduce the relative efficiency advantage

TABLE 5-7 Sulfur Content of Chinese Diesel Fuels (parts per million)

Sulfur Content	1995	1996	1997	1998
Minimum	300	300	300	300
Maximum	3,300	3,500	3,700	3,500

NOTE: Average sulfur content of refinery's diesel fuel pool.

SOURCE: Calculations by Qiu Yansheng, Research Institute of Petroleum Processing.

of the diesel engine over a spark ignition internal combustion engine. Fuel costs will not be changed significantly, but a large increase in diesel demand will require some modifications to existing refineries and their operations.

SINOPEC distributes diesel fuel in cities on the mainland, including the broad suburban areas. The diesel fuel grade depends on the specific contract between the users and the fuel suppliers. Some cities are requiring specifications that exceed the national standards. In principle, CNPC will follow SINOPEC in updating the performance of domestic fuels to meet the needs of the marketplace, with the advantage of having the lowest-sulfur crude oil source.

Diesel Fuel Specifications

The original state specification of diesel fuel, GB 252-1994, was replaced by GB 252-2000 in January 2002 (China State Bureau of Quality and Technical Supervision, 1994, 2000). GB 252-1994 covers three grades of diesel—regular, premium and super—based on sulfur content (see Table 5-8). The maximum sulfur contents of the three grades of diesel fuels are 10,000, 5,000, and 2,000 parts per million (ppm), respectively. The minimum cetane number limit is 45, except for the diesel fuels made from naphthenic or paraffin-naphthenic crude oils, as well as the diesel fuels containing catalytic cracking components, which have a minimum cetane number limit of 40.

GB 252-2000 includes one grade with a maximum sulfur content of 2,000 ppm. The minimum cetane number limit is 45, except for the diesel fuels made from naphthenic or paraffin-naphthenic crude oils, which have a minimum cetane number limit of 40. These specifications imply that the cetane number of the diesel fuels containing catalytic cracking components has a minimum limit of 45 rather than 40. In 2000 SINOPEC issued a city diesel fuel specification, Q/SHR 008-2000. In this industrial specification, the maximum sulfur content limit is 300 ppm, and the minimum cetane number limit is 50 without exception. The further development of diesel fuel specifications will focus on reducing maximum sulfur content to 50 ppm, which is the level needed to introduce Euro IV and Euro V emissions limits.[3] Indeed, China has a short-term schedule for improvements in diesel fuel (see Table 5-9).

As the Chinese economy develops, the demand for petroleum products will grow dramatically, and the shortage of crude oil will grow as

[3] As noted earlier, efforts are under way to lower this level to 10 ppm to enable some of the advanced aftertreatment technologies required to achieve the European standards.

TABLE 5-8 China's Diesel Fuel Specifications

	Diesel fuel specification GB 252-1994						Diesel fuel specification GB 252-2000						
Brand number	10	0	−10	−20	−35	−50	10	5	0	−10	−20	−35	−50
Solidifying point, °C max.	10	0	−10	−20	−35	−50	10	5	0	−10	−20	−35	−50
CFPP, °C max.	12	4	− 5	−14	−29	−44	12	8	4	− 5	−14	−29	−44
FP (PM), °C min.	65			60	45		55					45	
Cetane number* min.	45						45						
Distillation, °C: 50% vol. recovered at max., 90% vol. recovered at max., 95% vol. recovered at max.	300, 355, 365						300, 355, 365						
Density at 20°C, kg/m^3	Report						Report						
Viscosity at 20°C, mm^2/s	3.0–8.0			2.5–8.0	1.8–7.0		3.0–8.0			2.5–8.0		1.8–7.0	
Particulates	Nil						Nil						
Copper corrosion rating (50°C, 3 hr) max.	1						1						
Ash, % by mass. max.	0.01 (super and premium); 0.02 (regular)						0.01						
Carbon residue on 10% distillation residue, % by mass. max.	0.3 (0.4 for some grades of regular)						0.3						
Acidity, mg KOH/100 ml max.	10 (regular); 7 (premium); 5 (super)						7						
Water, % by vol. max.	Trace						Trace						
Sulfur, ppm max.	10,000 (regular); 5,000 (premium); 2,000 (super)						2,000						

Sulfur of mercaptan, % by mass max.	0.01(super and premium); no requirement for regular	
Color number max.	3.5 (super and premium); no requirement for regular	3.5
Iodine number, g I/100 g max.	6 (super); no requirement for premium or regular	
Oxidation stability (insoluble), mg/100 ml max.	2.0 (premium); no requirement for super or regular	2.5
Existent gums, mg/100 ml max.	70 (regular); no requirement for super or premium	

NOTE: Blanks indicate that no specification exists for these categories. CFPP = cold filter plugging point; FP (PM) = flash point (Pensky-Martens test); KOH = potassium hydroxide; I = iodine; kg/m^3 = kilograms per cubic meter; hr = hour; mm^2/s = square millimeters per second; mg = milligram; ml = milliliter; g = gram.

SOURCE: China State Bureau of Quality and Technical Supervision (1994, 2000).

TABLE 5-9 Planned Sulfur Standards, China

Timing	Nationwide	City
2001	Sulfur standard for light diesel fuels: 2,000 ppm (super); 5,000 ppm (premium); 10,000 ppm (regular)	SINOPEC provides cities with 500 ppm sulfur diesel fuel.
June 1, 2002	Sulfur standard for light diesel fuels: 2,000 ppm	
End of 2003 or early 2004	Sulfur standard for city diesel fuel: 500 ppm	SINOPEC will supply 300 ppm sulfur diesel fuel to metropolitan areas.

NOTE: ppm = parts per million. SOURCE: Calculations by Qiu Yansheng, Research Institute of Petroleum Processing.

well. As noted earlier, by 2010 about 100 MMT in crude oil will have to be imported annually to make up the shortage in the domestic supply. Meanwhile, China is improving the quality of gasoline and diesel fuel for meeting the more stringent requirements of environmental protection and the demands of the automotive industry. The main goals will be to reduce the olefin content of gasoline and the sulfur content of gasoline and diesel.

ALTERNATIVE FUEL POSSIBILITIES FOR THE FUTURE

China is already using some fuels other than gasoline and diesel in the transportation sector. Table 5-10 shows the shares of energy consumption by fuel source in China in 2000, both the total and for the transportation sector. China's main energy sources are shown in Table 5-11.

TABLE 5-10 Energy Consumption of China, 1997 and 2000

	Coal	Petroleum	Natural Gas	Electricity
Total energy share, % (1997)	76.2	19.7	1.8	2.2
Total energy share, % (2000)	67.0	23.6	2.5	6.9[a]
Transportation energy share, % (2000)	6.0	69.0[b]	n.a.	25.0

[a] Hydropower.
[b] Including natural gas.

NOTE: n.a. = not applicable.

SOURCES: Zhang (2001); Xinhua News Agency (2001).

TABLE 5-11 Chinese Energy Reserves

	Petroleum	Natural Gas	Coal
Proven recoverable reserve	3.3 billion metric tons (2000)	1.37 trillion cubic meters (2000)	114.5 billion metric tons (2000)
Annual output	162 million metric tons (2000)	27.7 billion cubic meters (2000)	1,045 million metric tons (1999)
Years at present rate	21	49	110

NOTE: "Years at present rate" assumes that the increased rate of energy resource use will match the increased rate of discovery.

SOURCES: Radler (2000); Zhu and Song (2000); Chen (2001).

Natural gas use rates are presently low, but major development of the natural gas infrastructure is under way to allow substitution of natural gas for coal, primarily for domestic uses and for power generation, in polluted urban areas. Natural gas is a clean fuel; nitrogen oxides (NO_x) are the primary emission of concern. Its carbon emissions are about 25 percent lower than those for gasoline, but whether it has net GHG benefits depends on how much methane is leaked. Methane is a more powerful greenhouse gas than carbon dioxide (CO_2) by about a factor of 20, averaged over a 100-year period, so a 1–2 percent leakage of methane can offset the lower carbon advantage of natural gas. High-pressure gas is less dense than gasoline and requires a pressurized cylindrical storage vessel, so fuel storage volume impinges to some extent on the space inside a vehicle. Liquefied natural gas (LNG), another alternative, involves liquid storage at about –160°C (–260°F) and a pressure just a little above atmospheric. LNG has a lower density than gasoline, requires about a 10 percent energy penalty for liquefaction, and must be stored in a sophisticated insulated container to minimize boil-off, so it is probably not as attractive an option as high-pressure gas storage for passenger vehicles. It is more attractive for fleet use where vehicles operate consistently from a central location.

Natural gas is now being used in about 110,000 vehicles in twelve cities in China, and about two-thirds of these vehicles are fitted with dual-fuel capability. Although natural gas spark ignition engines still emit nitrogen oxides, they are cleaner than diesel engines when used in urban taxis and buses. It is likely that the use of natural gas for transportation will continue to expand, but it will remain a minor part of the total energy used in the transportation sector. Indeed, after 2010 China will likely have to import natural gas to meet the growth in its domestic demand, so that large-scale substitution of natural gas for petroleum in the transportation sector does not appear to be a suitable answer to China's concerns about foreign oil dependence. However, natural gas is a clean fuel that has advantages for use in polluted urban areas.

Liquefied petroleum gas (LPG), a mixture of propane and butane, is a byproduct of oil and gas production from refineries. It is a useful transportation fuel, but its availability is limited.

Because *biofuels* are derived from solar energy, they are a dispersed energy source and consume a great deal of agricultural land (see Chapter 4). The Brazilian automobile fleet was fueled by biofuels for many years, but in light of current oil prices, dedicated ethanol vehicles have been largely abandoned. Biofuels are likely to continue to be a small part of the overall transportation energy mix, especially where they are used as additives or to fuel a small local vehicle fleet.

Methanol, yet another alternative vehicle fuel in China, is being used in Shanxi Province in some commercial vehicles. There, the annual output of fuel methanol from coal is expected to reach 3.8 MMT by 2005.

China's *coal-bed methane (CBM)* resources are substantial. In fact, China may have up to a third of total worldwide CBM resources, estimated to be in the range of 85–262 trillion cubic meters (m^3). Some CBM is recovered commercially in the United States.

China's chemical processing industry has already adopted coal gasification technology. Because conventional natural gas resources are scarce in China, 70 percent of its ammonia (NH_3) production in 1990 was based on the gasification of some 37 MMT of coal. There is also great potential to tap the large reservoir of coal-bed methane—by using the carbon dioxide produced as a byproduct of the manufacture of ammonia from coal to stimulate the recovery of methane from deep, unmineable coal-bed formations.

Long-term energy carrier choices that ultimately are of interest are either *electricity* or *hydrogen*. As discussed in Chapter 3, electricity is not a competitive energy source for personal vehicle transportation today because of the poor performance and high cost of batteries for onboard vehicle energy storage. Because electricity can be produced from coal, hydropower, or nuclear power, it has the inherent potential to be a large-scale energy source in China for the transportation sector. If future technological developments lead to batteries that provide cost-efficient mobile energy storage, electric propulsion can offer a significant alternative for consideration. But even if this comes to pass, the GHG issues associated with the production of electricity from coal must be addressed. Hydrogen, an alternative energy carrier to electricity, has the advantage of being somewhat easier to store than electrons in a battery. However, like electricity, hydrogen has to be produced from a primary energy source. The commercial source for hydrogen production today is natural gas. Limitations on the availability of natural gas will apply equally to its use to produce hydrogen as a transportation fuel. If it were economically available and if the infrastructure existed to make it geographically available, it would be the best fuel for fuel cell vehicles.

Indeed, the development and promotion of fuel cell vehicles in China will depend on China's source of hydrogen. One possibility is China's huge reservoir of coal and coal-bed methane, if affordable hydrogen production can be accomplished with the minimal emission of greenhouse gases. According to Tables 5-10 and 5-11, coal is the dominant source of China's energy supply, but it contributes only about 6 percent of total energy use in the transportation sector.

The use of coal gasification to produce synthetic liquid fuels or hydrogen for transportation results in energy inefficiencies and the increased generation of greenhouse gases. Although large quantities of carbon dioxide are generated in the manufacture of hydrogen, CO_2 emissions could be reduced drastically if CO_2 sequestration technology becomes available. The world's scientific community is increasingly confident that the sequestration of a significant fraction of global CO_2 production from the use of fossil fuels over the next several centuries may be feasible, especially in light of new understanding of the potential for sequestration in geological reservoirs—that is, depleted oil and gas fields, deep saline aquifers, and deep beds of unmineable coal. However, these technologies are still in the early stages of development and are likely to involve significant added costs.

REFERENCES

Chen M. 2001. Demand for crude oil and supply in China. In Proceedings of 2001 Conference on Crude Oil Information. SINOPEC Crude Oil Information Center, Beijing (in Chinese).

China State Bureau of Quality and Technical Supervision. 1994. Chinese Light Diesel Fuel Specification GB 252-1994 (in Chinese).

———. 1999. Chinese Unleaded Gasoline Specification GB 17930-1999 (in Chinese).

———. 2000. Chinese Light Diesel Fuel Specification GB 252-2000 (in Chinese).

Histon, P. 2001. Analysis and Prediction of Clean Fuels in Asia. British Petroleum Technical Centre, Sunbury-on-Thames, UK.

Owen, K., and T. Coley. 1995. Automotive Fuels Reference Book. 2d ed. Warrendale, Pa.: Society of Automotive Engineers.

Radler, M. 2000. World crude and natural gas reserves rebound in 2000. Oil and Gas Journal 98(51):121–123.

Ruan Y. 2001. Production status, problems and countermeasures of Chinese gasoline and diesel oil. Petroleum Products Application Research 19(6):10–14 (in Chinese).

Schafer, A. 1998.The global demand for motorized mobility. Transportation Research A 32(6):455–477.

Stell, J. 2000. Worldwide refining survey. Oil and Gas Journal 98(51):66–68.

U.S. Department of Energy (DOE). 1999. International Energy Outlook. Energy Information Administration, DOE, Washington, D.C.

Xinhua News Agency. 2001. China Petrochemical News. November 23 (in Chinese).

Yang J., Song W., Wang L., and Wang J. 1997. Analysis for supply and demand of world crude resource and trend of China's petroleum industry. In Proceedings of 15th World Petroleum Congress, Beijing, 1997 (in Chinese).

Zhang L. 2001. On China's energy security. International Petroleum Economics 9(3):10–14 (in Chinese).

Zhu X. and Song W. 2000. Strategic significance of development of coal bed gas: Importance, necessity, and urgency. Chinese Energy (10):8–12 (in Chinese).

6

Societal Effects of Potential Motorization Pathways

In many countries the automobile has provided benefits that have boosted the quality of life: better health because of quicker access to medical care, more highly skilled workers because of better access to education, a greater choice of jobs because more are in reach of the average worker, better prices from regional retailers because more customers can go elsewhere, and more recreational opportunities because distant vacation venues are more accessible. In short, the automobile has increased equality through greater choice and liberty through the freedom of movement (NRC, 1999).

Yet the automobile also has had negative effects on quality of life, many of them environmental (NRC, 1997; Forman et al., 2002). By the 1950s the air pollution created by motor vehicle exhausts and by the refining of fuel for those vehicles had poisoned the air in some cities to such an extent that it was shortening life expectancy. In the United States a significant proportion of water pollution can be traced to motor fuels and the oils and salts that wash off roadways. The disposal of old automobiles also is a source of environmental problems. In some places millions of old tires fill entire valleys, and the disposal and reprocessing of used motor oil are a serious environmental problem.

Over the last 30–40 years some industrialized countries have made enormous progress in balancing environmental concerns with automobile usage. Air quality has improved through technology such as advanced engines, modified fuels, and catalytic converters. Highway agencies are required to collect the polluted runoff water from highways for treatment before letting it flow into rivers. New automobiles are designed so that many of their components can be recycled at the end of their useful lives.

Among the most critical aspects of transportation planning in the past have been those related to equity issues. Although technological improvements and capital investments in transportation have improved social welfare in general, they also have often widened the gap between those who can afford automobiles and those who cannot. To the extent that modern land use patterns are highly affected by the automobile, the result is that car ownership has become necessary to take full advantage of the commercial and residential opportunities and of the services that modern societies can provide. In the United States, the disadvantaged are primarily the elderly, the young, and the poor, yet today less than 10 percent of American families lack cars. In China, however, large urban populations will not have cars for many years, and for a long period car owners and non–car owners will both be large parts of the population. In some countries these populations live in separate communities, sometimes distinguishing urban and suburban areas.

This chapter explores how rapid motorization will affect the quality of life and livelihood of the Chinese people. In view of the complexity of the changing social and economic structure in China, it is clear that isolating the changes arising from motorization alone will be very difficult. Some 10 years of a 15 percent a year rise in motorization in general and a more rapid increase in the number of automobiles have had a very large impact on China. Urban congestion, air pollution, changes in urban transport modal shares, massive construction of transportation facilities, and decentralization of residence and employment are the principal consequences (Midgley, 1994).

The earliest and greatest effects of motorization will appear in the cities (see Chapter 2). For one thing, most of the new vehicles will appear in urban areas because urban incomes are higher than those elsewhere in China and because they will no doubt continue rising the fastest. Furthermore, the cities have an especially delicate ecology of space use—very high residential densities and a low proportion of space dedicated to streets (around 10 percent)—designed as they were by planners who, before the late 1980s, had no reason to believe there was any need to provide for significant numbers of private motor vehicles. As a result, Chinese cities are subject to great dynamic forces as people struggle to travel in increasingly congested conditions while they seek opportunities at the urban periphery made accessible by increased car ownership.

Because these issues will arise within the context of a rapidly changing urban structure, any examination of the effects of motorization requires a careful look at the process of change itself.

EFFECTS OF THE MOTORIZATION PROCESS

Motorization is likely to have severe interim effects before the automobile reaches an accommodation with its surroundings. For example, residences typically shift to the suburbs more rapidly than employment, creating an interim period during which employees have long, congested radial trips to work. The current level of motorization is less important than its rate of increase, because it is the rapid increase of motorization that leaves the urban structure, the infrastructure network, and the access system out of sync with one another. For example, today developed countries with over 900 vehicles per 1,000 persons have reached a limited kind of equilibrium with the automobile (at least for congestion in most localities), yet that transition in the United States took more than 60 years, with some changes still under way. By contrast, developing countries with only a tenth of that level of motorization are tending to experience greater problems than the developed countries because of the rapid shift to motorization.

The growth rate of motor vehicles in Chinese cities has mirrored China's rate of economic growth, and with the accelerated economic growth of the early 1990s, the growth rate of motorization became more rapid. In the city of Nanjing, for example, between 1985 and 1990 the total number of motor vehicles increased by around 50,000, whereas between 1995 and 2000 the number increased by around 130,000, reflecting the city's economic growth. In the city of Shanghai the growth rate of automobiles was slow until the 1980s; the average annual increment was 13,000. But during the 1990s the growth rate more than doubled, with annual increments of between 30,000 and 50,000. In Wuhan the number of vehicles increased 7 percent from 1998 to 1999, but from 2000 to 2001 it increased 27 percent. Even though annual changes can stem from many factors, China clearly has entered the phase of rapid motor vehicle growth. And with it, problems of urban transport have increased rather than diminished.

Another problem is that of equity among China's regions. The industrial cities of eastern China have wealthier corporations and more people who can afford cars than the western provinces. Motorization will benefit the wealthier families of the eastern cities, modernizing their lifestyle, notwithstanding the additional congestion and air quality problems caused by the larger vehicle fleet. As a result, the difference in quality of life between the industrialized east and the rest of the country will be magnified, exacerbating the regional imbalance that has been a persistent problem for China (and for many other countries).

The Spatial Decentralization of the Major Chinese Cities

No issue is more central to motorization than the shift of people and industry from urban centers to outlying areas—that is, spatial decentralization. As one of the most visible consequences of motorization in the medium term, it will change the appearance of the cities and the lives of their inhabitants. Moreover, it will affect nearly every other measure of the social consequences of motorization such as fuel consumption, air pollution, land absorption, viability of alternative modes of transportation, and congestion.

Municipal governments have taken some deliberate actions to reduce the density of the cities in order to reduce congestion and to adapt contemporary lifestyles and associated technologies. To some extent these actions will mitigate some of the congestion in the short term. In the immediate future, the problem will be to accommodate the car in areas of the city not developed with cars in mind.

The view of many Chinese authorities that population densities in large Chinese cities should decline has merit. But the added factor of rapid motorization is likely to accelerate the decentralization of jobs and residences beyond planned levels and should be a cause for concern. For example, such decentralization could jeopardize the feasibility of public transport (which cannot serve low-density areas effectively). Expanded road networks could increase congestion at key intown links, which would be hard to correct without a high cost and severe disruption. Decentralization also would tend to divide the population into high-income (car-owning) and low-income (carless) localities, threatening the Chinese tradition of socially integrated communities (locating according to work group membership). And because decentralization means people will drive farther, emissions of pollutants and greenhouse gases will rise and more fuel will be consumed.

Another concern arising from decentralization is the fate of China's agricultural land. By world standards, China's amount of agricultural land per capita is very low—0.1 hectares (ha) per inhabitant—and the average rural population density is a high 6.85 persons per hectare or 685 persons per square kilometer (World Bank, 2001). The rapid growth of the cities will absorb large amounts of agricultural land, because most cities are located, for historical reasons, in rich agricultural areas. For example, the rapidly motorizing cities of Shanghai, Nanjing, and Wuhan are located in the rich Yangtze Valley.[1] Authorities in Guangdong Province have expressed concern about the absorption of agricultural land by urbanization in the Pearl River Delta.

The green movement in Europe and those who resist the increasing use of cars in the United States often appear to assume that the automo-

[1] A brief calculation indicates that if Shanghai's population in 1990s were distributed at the density of the New York City metropolitan area, Shanghai would be about eight times its actual geographic size (www.demographia.com; Newman and Kenworthy, 1989; State Statistical Bureau, 1991).

bile is the only cause of spatial decentralization. Yet other forces work toward decentralization in every country, including China. After World War II, for example, explosive decentralization in the United States was driven by the easily available home mortgages on newly constructed suburban houses. Urban policies intended to limit development often focus exclusively on resisting vehicle use or road investments, but other matters should receive attention.

Government action drives decentralization in many ways. For example, new urban highways and public transport systems reaching toward the outer fringe help to open new peripheral areas to settlement. The Chinese government recently set national development standards that require more parking space, wider streets, and increased per capita living floor space. The new government guideline, enacted through the Construction Department, calls for an average density of 80–100 square meters —that is, 12,500–10,000 persons per square kilometer—including all land uses averaged over the entire metropolitan area (Construction Department of China, 2001). The larger cities are to aim at the less dense figure. This goal is to be reached gradually while new development at the urban periphery is designated for higher densities and the continued reconstruction of central cities reduces densities in the core. This would have the effect of driving development toward the larger land parcels available in the urban periphery. Furthermore, Chinese home construction, which is no longer the exclusive responsibility of the state-owned enterprises, is increasingly being carried out by developers, who are likely to seek economies of scale by building larger housing projects. They tend to build at the periphery of cities where larger land parcels are available, prices are lower, and land acquisition and consolidation are easier. Developers also seek large parcels because the government encourages them to provide municipal services. That factor further drives settlements to the outer edge of urban areas (Ping and Murie, 1999).

Previously, when land was assigned to state-owned enterprises the value of the land did not vary with distance from the city center. No money was transferred when the land was assigned, and the distance from the enterprise's own prior workplaces and service facilities primarily determined its value to the work group. Since 1987 municipalities have been allowed to lease land and to assign monetary values to the land parcels as a basis for levying charges. The land value is determined, in part, by the distance from the center of the city, thereby encouraging parcel seekers to choose peripheral locations for the lowest prices. This system of designating land value has been important in redirecting urban development toward the periphery. Meanwhile, many of the state-owned enterprises are finding that the land they occupy in the city center has greatly increased in value, but at no benefit to their operations. They are therefore inclined to redevelop this land into central business district uses and move their own operations to the periphery. The outcome, then, is that the redevelopment prompts the sponsoring enterprise to leave the city, as well as the worker population that resided at the former site of the enterprise (Rodrigue, 1994).

Perhaps reflecting an international trend, new industrial plants, universities, and other large establishments in China now tend to locate in very large, campus-like settings, where the buildings themselves occupy very little of the land. Such a trend is another force behind decentralization, because such large parcels are found only at the peripheries of cities. Transnational companies attracted by China's accession to the World Trade Organization (WTO) will no doubt adopt and accelerate this trend. Many of the new industrial plants in the Pearl River Delta are dramatic examples of this movement.

When municipalities purchase rural land and redesignate it as "urban," they are able to sell it to developers for much more than its cost. The municipalities therefore have an inducement to annex land in large quantities in order to raise revenue. Indeed, when land is leased long-term to developers, taxes (infrastructure fees) are charged for the entire length of the lease. Thus the municipality gains no further revenue from the land during the length of the lease, even though the value of the parcel may rise greatly over time. As a result, municipalities are motivated to continue converting rural land to urban in order to have continued land-based revenue.

At the same time, municipal planning offices are actively fostering concentrated dispersion. Fourteen "growth centers" (existing expandable small settlements) are being planned for Beijing, and 11 new towns and 20 external growth centers are planned for Shanghai. These centers, to be situated well beyond the urban periphery, are often designed for very low densities. Commercial buildings and apartments will be placed on large parcels, and the centers will include large central parks and wide, multi-lane local streets.

Trends in the city of Guangzhou are representative of this process. Figure 6.1 shows the difference between land development initiatives before and after the land market reform of 1987. Note how the reform has concentrated new development in both the city center and peripheral concentrations 21 and 29 kilometers from the center where the government has encouraged growth (Wu and Yeh, 1999).

Another incipient, powerful force behind decentralization is the popularity of portable modes of electronic communication. It permits people to substitute electronic connections and phone calls for trips. Currently in China, there are about 7 conventional telephones for every 100 people (29 per 100 in Shanghai) and long-distance calls are very expensive (World

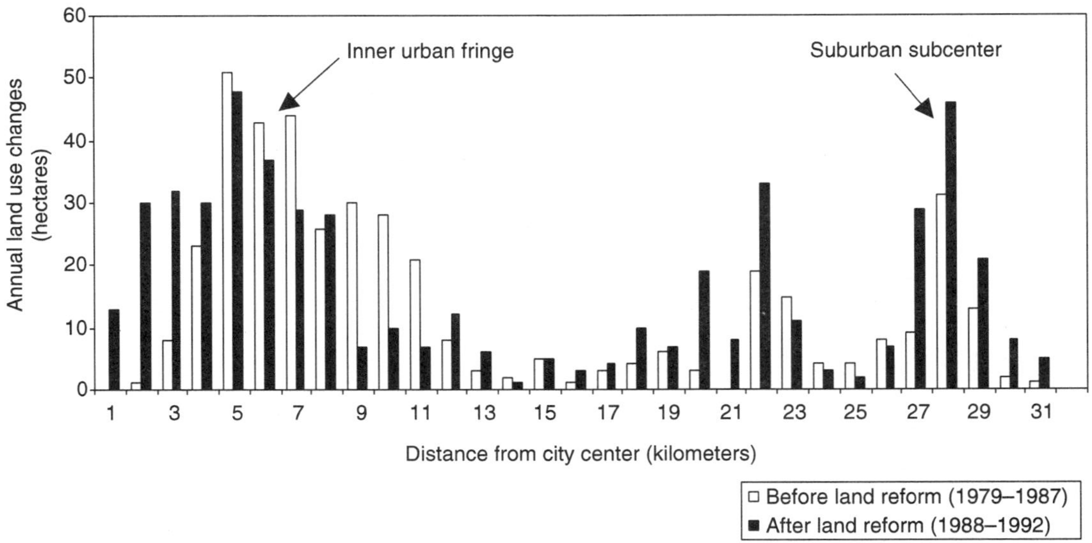

FIGURE 6-1 Distribution of land use changes outward from city center of Guangzhou, before and after land reform of 1987. SOURCE: Wu and Yeh (1999).

Bank, 2001). As a result, this traditional mode of communication has contributed little to decentralization. Since the 1990s, however, many communications technologies have become available at once, such as the cell phone, general Internet use, and various forms of electronic access (e.g., corporate databases). It is important to note that Chinese culture is showing great flexibility in adapting to the change, unlike countries in the West, where these communications technologies had to tug on long-standing economic organizations with their well-defended vested interests.

While there is no doubt that these forces will produce more livable environments, access to less expensive land and better housing for those able to enjoy them, they also produce a strong momentum toward decentralization. Although motorization is not the only cause of urban decentralization, transportation policy must serve as a very powerful lever to guide urban development into the motor age in a rational manner. Productive decentralization requires careful orchestration of the forces driving it.

Because of all the forces cited, Chinese cities are expanding very rapidly, yet a close look at various world cities with different urban development policies and different densities suggests that the cities of China have choices in this matter. Currently, Shanghai's density is 16,378 persons per square kilometer.[2] But across the world urban densities vary greatly. What explains the differences? Of the many factors, the most important, aside from geography, are income/motorization, regulation of land development, and transportation policy.

Somc citics have been influenced by special factors and so are of little interest here. Hong Kong, for example, possibly the world's densest major city at 28,405 persons per square kilometer, was once an isolated colony without room to expand geographically. With its high incomes, expensive high buildings were feasible for apartments, and cars were extremely expensive and stringently regulated.

Seoul, at 23,908 persons per square kilometer, is nearly as dense as Hong Kong. South Korea is one of the higher-income Asian countries, with a history of highly planned economic development during which motorization was firmly discouraged by very high auto purchase and use prices. Seoul still has very stringent land use regulations, including provisions that protect a green belt that contains metropolitan development and retains urban densities

Jakarta, at 17,056 persons per square kilometer, and Bangkok, at 14,955, have roughly the same density as Shanghai, but their densities result from the ongoing operation of urban land markets, whereas Shanghai's results from a half-century of centralized planning. Although

---

[2] All the density figures provided here were obtained from www.demographia .com.

both Bangkok and Jakarta have large numbers of low-income people living at high densities, neither has imposed significant controls over urban expansion or motorization, nor experienced the sudden, rapid changes in political economy that are partly responsible for Chinese decentralization.

Kuala Lumpur, at 5,693 persons per square kilometer, stands in stark contrast to the cities just described. It is characterized by very rapid economic development, a very high level of motorization, and no effort to contain urban development densities. In fact multiple forces encourage very rapid decentralization, especially the government's tendency to give very large land grants to institutions. At the same time, traffic congestion in the city center is intense. If economic development continues apace and if strong preferences for decentralization-forcing policies are sustained, the prospective densities for Chinese cities will approximate those of Kuala Lumpur.

Tokyo (at 7,099 persons per square kilometer) is a high-income city with an array of strong urban development containment policies. Holding development to this density as Japanese income levels rise requires, and receives, strong public support. Auto ownership in Tokyo is very expensive by world standards, and so within the city a very high percentage of trips are made by public transit.

New York and Chicago, at 2,086 and 1,653 persons per square kilometer, respectively, are friendlier to motorization and decentralization. Auto ownership is about twice as high as in Tokyo, residential densities are low, and the costs of gasoline and the other aspects of car ownership and use are much lower in both New York and Chicago. Land zoning is applied to avoid incompatible land uses, but it is generally not used to maintain a density of development. If anything, it is used to keep densities below specified maxima in fundamentally low density environments.

Although these brief sketches leave out a great deal that would further explain urban density, they do show the most basic conditions and options for different urban structures. The Chinese city is at the beginning of a new era. It has the strength of governance that would enable it to make telling choices.

Challenges for the Nonmotorized Population

During the 1990s many citizens of Shanghai with medium and low incomes moved to the peripheral areas because of the demolition of houses by municipal engineering projects. When the new residential quarters were built, public transport lines were not yet in operation. As a result, low-priced mopeds became the transport of choice, until recent years when restrictions were placed on ownership because of safety and air pollution considerations.

In view of the current income profiles of urban populations and the expectation that low-income people will continue to migrate from agricultural areas, it is expected that the nonmotorized urban population will continue to be large. The migrant "floating" population tends to be largely concentrated in informal settlements near suburban work opportunities rather than in city centers, which are too crowded to absorb them.[3] These migrants will join the urban populations who presently rely on bicycles, public transportation, or their own feet for transportation to work.

Workers without cars who are faced with longer distances to work because of decentralization also will face longer transit trips, and the transit itself will be significantly slowed by the heavier auto congestion. New auto owners face congestion too, but they can choose less-congested, circuitous routes or different destinations. Buses, however, typically must use the main arteries, the most congested part of the network. Ironically, the greater the congestion, the greater the advantages of auto ownership. As a result, municipalities will find it more important to examine the possibilities of retaining bicycle transport where possible and providing housing and services in the localities of employment (Midgley, 1994).

CONSTRUCTION OF TRANSPORT FACILITIES IN CHINESE CITIES: THE CASE OF SHANGHAI

The evolving pattern of streets and highways in major Chinese cities has a hierarchy that is similar to that found in other countries. The narrow streets in the residential parts of older cities are interspersed with arterials that carry longer-distance traffic. Many cities have built limited-access urban expressways. The outer, newer parts of metropolitan areas are lower in density and devote more land to street systems. Although there are significant differences among Chinese cities, with Beijing perhaps the most endowed with ample central arteries, the general situation is similar. The arterials, and in many instances the urban expressways, are badly congested during much of the day. Nearly all streets that carry through traffic (rather than simply providing access to nearby property) are now congested and sure to become much more so with currently rapid motorization rates.

---

[3] Everyone in China has a registered permanent residence that is assigned at birth. Citizens who move to a location where they do not have a local registered permanent residence may not remain in that place unless they obtain a license from the police every half-year (the household registry system). "Floating population" refers to the people who move from one place to another, usually staying less than a half-year with no official change in their registered permanent residence.

The Chinese government's Code for Urban Road Planning and Design is attempting to relieve this situations by enacting a guideline for cities that calls for "road coverage" of 8–15 percent (from the current 10–12 percent) for smaller cities, and 15–20 percent for cities with a population of over 2 million. These figures, intended as averages over an entire metropolitan area, would include the running lanes (at about 10 percent); parallel lanes for two-wheeled vehicles, parking, and pedestrians; and park areas (Technical Supervision Bureau and Construction Department of China, 1995).

During the 1990s Chinese cities, especially Shanghai, made forceful efforts to keep up with motorization rates. Over the decade, Shanghai increased its road length by 40 percent and its public transportation routes by 30 percent, and three subway lines were developed for a total of 65 km. At the same time, the number of motor vehicles increased two and a half times, to more than 700,000. The public transportation services and the traffic management program have improved dramatically. For example, the investment in the city's transportation systems accounted for 2.6 percent of Shanghai's gross domestic product (GDP) in 1999, a very high rate by world standards. The investment program has emphasized public transport, especially subways, but it also has financed the development of roads, especially in the outer parts of the metropolitan area (see Appendix B).

Despite these investments, many of the congestion problems expected from rapid motorization are already present. For example, the Inner Ring Road of Shanghai, completed in 1995, is already congested with traffic. Average vehicle speeds in Shanghai are about 15 kilometers per hour (kph) and rarely exceed 30 kph. Average transit service speeds are 10–16 kph. In 1995 the average travel time for an auto trip was 55 minutes, one of the highest average trip times among the world's cities.

As noted, Shanghai has attempted to keep up with the infrastructure requirements for urban transportation. From 1991 to 1996 the city spent approximately RMB83 billion ($10 billion) on projects that included two major bridges, a tunnel under the Huangpu River, an inner ring road, and the first line of its new subway system. In 2000 a new development plan took effect that calls for numerous improvements by 2020, including 200 km of rail, six elevated busways, and 650 km of divided highway in suburban areas (of which 520 km will be new highway), three new river crossing facilities, and water and air regional transport facilities.

At the same time Shanghai and other cities have adopted various traffic management strategies to control the rising congestion. These include stringent regulations on the time, day, and locality of the use of trucks. Various fees also have been imposed on cars, at purchase and on use. Moreover, the city has made forceful efforts to limit the use of two-wheeled motor vehicles, such as excluding those not registered in the jurisdiction and not accepting new registrations. Efforts to restrict the use of bicycles are aimed at avoiding the conflicts that occur among two- and four-wheeled vehicles in parallel lanes.

Chinese cities are caught between the desire to promote auto motorization and the problem of accommodating all the new cars. Agencies in Shanghai have even proposed ceilings on the number of cars permitted within the city (Shanghai Study Team, 2001). The City Transport Administration Bureau has called for a limit of 1–1.3 million cars until 2020. But even with such a limit, it would be a considerable challenge to provide the necessary infrastructure in the coming years. The experience of other countries (see Chapter 2) demonstrates that, while the number of urban vehicles increases in proportion to income, urban road length does not. Road building in cities is expensive and politically sensitive because of difficulties encountered in acquiring land, relocating businesses and households, displacing utility pipes and wires, and dealing with the confined conditions of construction. Neighborhood interests also may resist construction—examples of such resistance are already evident in Chinese cities.

IMPACT OF MOTORIZATION ON URBAN TRAFFIC MANAGEMENT REQUIREMENTS

In general, most rapidly motorizing Chinese cities can expect to experience the following conditions and actions, based on the experience of other cities in the motorizing world:

- The need to construct urban highways will continue, perhaps by concession to private firms and financed by tolls in many cases. In particular, congested links will be reconstructed to increase their capacity.
- Management of the parking stock is required. The investment in private parking spaces near city centers can be expected to occur when the value for parking becomes competitive with that for other uses. This construction will make congestion worse by encouraging arrivals. Employers will likely provide more parking for their employees, adding to the traffic pressure.
- A demand for the construction of public transit facilities will emerge, principally for people still without cars in an increasingly auto-oriented city. (Worldwide experience is that few trip makers will change from auto to transit in the short run.)
- Employers will increasingly provide private bus services for their employees, possibly prompting regulatory requirements because the services will compete with public transit.
- The demand for limitations on bicycle use will increase, especially

in the city center, even though for the short trips within the congested city center bicycles are the most effective alternative. This problem is generating an intense debate between those attempting to facilitate traffic and those defending bicycles as effective and environmentally benign transport, especially for lower-income people.

- The demand for restrictions on the use of two-wheeled motor vehicles will increase, possibly ending in their termination in some cities.
- The use of trucks will have to be further restricted, tempered by the need to maintain efficiency of the economy.
- As the number of autos increases, calls will emerge for limitations on their use. So far, other modes of transportation have been controlled by direct traffic management, whereas autos have been controlled principally by pricing methods. Intensified congestion pricing of cars, such as time or area licensing, bridge and tunnel tolls, or parking charges, might be considered, but it is likely that cities will consider direct-use limitations unless there is intervention at the national level.

Travel Demand Management

When the demand for travel increases rapidly, travel demand management takes on special importance. In China such management techniques have already been developed and applied to a high degree. They include regulating trucking (such as requiring night delivery), limiting the number of days trucks can be used, and limiting areas of the city to trucks with specific uses. These techniques also include special lanes for two-wheeled vehicles, restriction of vehicles in some areas to those with local registration, lower parking fees for local vehicles, and various pricing strategies for the purchase and registration of cars. Especially stringent use of these technologies explains, for example, the slower growth of motorization in Shanghai than in other Chinese cities.

Chinese traffic planners could apply other high-yield options that have been used with effect elsewhere. For example, some cities have restricted the entry of traffic into downtown areas or other dense destination points by the use of special tolls or permits, complemented by parking facilities provided in quantity and price to limit downtown congestion. One of the first to employ this method was Singapore, where a licensing scheme succeeded in reducing auto entries into the downtown area by some 45 percent. Singapore has now gone on to citywide congestion pricing. Other cities that have enacted charges for entering the city center include Rome, Oslo, Bergen, and Amsterdam, and such charges also have been discussed seriously in London and in Tokyo (where seven separate districts are proposed for paid entry).

Still other cities have restricted entry into the city center physically by creating pedestrian ways through which motor vehicles cannot drive or cross. This technique, which compartmentalizes the traffic approaching the city center, is used in hundreds of European cities—among the best known are Munich and Göteborg—and in Seoul. In China, Shanghai and Beijing have pedestrian ways that are clearly popular with shoppers and strollers. Also in many cities in the developing world, even in the absence of regulations the entry of vehicles into central streets is effectively forbidden by the volumes of pedestrian traffic they carry.

The use of advanced electronics for driver advisory functions and incident detection in an intelligent transportation system (ITS) is a potentially important way of increasing road capacity. These techniques include electronic toll collection and real-time electronic or radio reports on traffic conditions and parking availability that help drivers to select travel times and routes to avoid congestion. Systems that provide information on obstructive incidents (accidents and breakdowns) could be helpful; up to one-half of travel delays stem from these events in the United States. Panels that direct drivers to the available parking spaces reduce cruising in search of parking. And computerized programs that guide drivers to their destinations cut back on congestion by reducing the number of cars that get lost in the streets or take overly long routes. It is hoped that knowledge of traffic conditions through online reports will encourage drivers to avoid peak hours, or switch to fast transit on independent rights of way to shorten their travel times.

Car sharing is rising in importance in various parts of the world (Gakenheimer and DeLisi, 2000). Various car-sharing companies provide economical, short-term auto rentals to members of the system from numerous small stations located throughout a city. It might be useful in China as a means of enabling families to secure a car for occasional use without owning (or having to park) one. In Chinese cities, where there are other modes of transport more suited to various kinds of trips, such an arrangement would enable trip makers to save money and avoid some of the difficulties of car ownership. Car sharing is growing rapidly in Europe, where the leading company, which is in Switzerland, has 600 rental locations and 20,000 members. Successful systems also can be found in France and Germany, and car sharing is growing in popularity in Britain, Italy, and the United States.

Another innovation that has been growing very rapidly in acceptance is high-capacity bus transport. Buses operate on dedicated rights of way, with small stations where passengers pay the fare before entering the bus. The progress of each bus is assisted by sensors that change the signal lights to favor the bus as it approaches. The buses (or trolley-buses) have a large capacity (up to 210 passengers) and large doors that facilitate rapid entry and exit. Originally developed in the southern Brazilian city of

Curitiba, this system has been installed in Quito (Ecuador), Bogota (Colombia), and Dublin (Ireland), and is being readied for operation in many other cities. In China, the system is currently undergoing trials for a short-distance test line at Kunming (Tiwari, 2002). In New Delhi, city authorities have decided to undertake a 20 km test line. Although subways have a larger capacity and the advantage of not removing a lane at the surface, subway construction in Chinese cities may well be limited by the very high cost. High-capacity bus transit costs about $5 million (RMB42 million) per kilometer, rather than the $100 million (RMB830 million) per kilometer cost of rapid rail transit. Moreover, it appears to have a maximum capacity of about 45,000 passengers per direction per hour, whereas metro has a capacity of about 65,000 passengers per direction per hour and leaves the surface lanes free.

Finally, those planning expansion of public transport in China may wish to consider various advanced electronic means of improving the efficiency of public transit. These include the use of smart cards for fare payment and the adoption of other routing and scheduling tasks. In Japan electronic platooning of vehicles is being undertaken, enabling buses to travel in train-like groups.

Accidents

The accident fatality rate (per vehicle) in China is about 30 times higher than that in the United States. The rate may appear high for a generally well-disciplined society, but it is consistent with that of other countries with similar income levels, as reported in Chapter 2. It may be the result of the complex and changing mix of vehicles in the streets, the congestion that limits access of emergency vehicles, and the growing number of new, inexperienced drivers and pedestrians facing heavy vehicular traffic for the first time. Instruction for children in schools that enables them to cope with traffic as pedestrians and driver education in higher grades would be an important help.

Recent reports on traffic accidents in China[4] reveal that from 1990 to 1999 the number of accidents, the number of people injured, and the direct economic loss from accidents doubled and the number of people killed increased by 70 percent. But during that period the number of motor vehicles in China more than quadrupled, which means that the numbers per vehicle declined by half or more. These trends suggest that accident rates will decline on a per-vehicle basis as motorization becomes a

[4] China Auto Consulting (CAC), various reports.

more familiar part of the Chinese scene. Nevertheless, accidents are numerous, and as motorization grows the overall personal risk of accidents increases.

THE SOCIAL IMPACTS OF RAPID MOTORIZATION

Forecasts of adjustments in social behavior resulting from motorization are necessarily very speculative. Many are cultural changes observed in cities of other nations that have been through the motorization process. Each culture, being unique, does not go through exactly the same process of adjustment as others, yet the response has been so similar in cultures so different from one another that a closer look at cultural changes produced by motorization is worthwhile.

It is likely that among families above a certain threshold of income the availability and use of a car will become habitual. As a result, more young people will have access to a family car that they can learn to drive. The long and expensive requirements for learning to drive will ease, and the current dominance of men among driver's permit holders will give way to more equal participation by men and women.

Many of the changes in behavior will appear in the young drivers of the second motorized generation who were able to practice driving their parents' cars. It is not likely that a family headed by parents in their late fifties who get a new car will change its social practices very much. But new, younger drivers are apt to see the geography of their city differently. They are likely to drive a lot more than their auto-owning parents.

Like most Americans and Europeans, these younger people may well own a car before they own a house. In China's rapidly expanding land and housing markets, it is difficult to predict the long-term value of a house bought at a given price today. Although a car is a purchase that depreciates, it is likely to be seen as a safer one, based on a 10-year lifetime. This may mean that a young family will already own a car when preparing to choose a house to own—not the opposite. This observation is important, because it also means such a family may choose a house where the traffic is tolerable and there is room to park the car, frequently at the outer periphery of the city.

At the outer periphery, young families will have a larger set of choices and opportunities: a greater selection of jobs, a wider choice of schools and specialized studies, a wider selection of residential localities, and a wider range of friends and personal associates. They also will be able to comparison shop before buying, and they will be able to shop for food once a week at larger stores rather than daily at smaller, more expensive stores. The range of personal access will encourage the creation of large-scale shopping centers that will serve large metropolitan subregions. With the recently shortened workweek, people will take more vacations and travel greater distances for outings.

But these profound changes will exclude the major part of the population, who will not be able to acquire automobiles and are likely to be without them for a long time. As in most partly motorized countries, this problem is likely to be starker because of spatial separation. People with cars will seek low-density environments where the housing is less expensive (per square meter), while those without cars will remain in the inner areas. This separation will occur, ironically, through a reversal of the current residential pattern, in which the higher-income people, and probably the highest auto ownership, are in the centers of the cities. This reversal has in fact taken place in many cities of the world. Motorization turns cities inside out.

In summary, local communities in Chinese urban centers may likely become residence-based communities, rather than work-based—a change that entails a very different social organization. The division in society may become serious when the auto-owning group becomes a significant portion of the total population, so that the two groups become aware of each other as competitors with different lifestyles and personal opportunities. Municipal governments may well become increasingly concerned about this division in the society, as they have in other countries. Indeed, the situation puts special priority on maintaining an adequate public transport system and attempting to create cohesion in communities that no longer have a state-owned enterprise employer to assure that cohesion.

In the course of the evolution of motorization, citizen interest groups are likely to appear, and many will start nongovernmental organizations for the purpose of advancing their interests. These groups are likely to include auto owners, bicycle owners, and urban groups opposed to highway expansion. Auto owners would probably advocate substantial continuing budgets to improve the mobility of auto users, an interest resonating with the national government's promotion of the auto industry. They also would encourage local municipalities to continue high levels of investment in roads.

Considerable conflict may develop over the role of the bicycle; after all, most Chinese will not own cars for the foreseeable future. State statistics indicate that in China there are nearly two bicycles per family—an average about constant for all eight octiles of income level nationally. Although many bicycle trips will become impractical in the decentralizing city, bicycles will remain useful for many others. Municipalities will find it very difficult to provide public transit that is reasonably competitive with bicycle trips in the range of 10 km. It is therefore likely that organizations will emerge to advocate separate rights of way and other solutions.

Urban groups opposed to highway expansion are likely to take on the issues that will arise from the public's experience in grappling with the rapid construction of transportation facilities. These issues might include the invasion of natural forests and, especially, agricultural land by the extension of suburbs beyond current urban limits; the damage to air quality because of the increased number of vehicles; and the dislocation and the demolition of valued buildings. One case that has attracted considerable attention is the expansion of Ping An Ave. (Peace Ave.) in Beijing, where the proposal to demolish historic buildings in the course of neighborhood redevelopment has generated opposition from the community.

CONCLUSION

This chapter has attempted to combine the emerging evidence of new motorization-impelled behavior in Chinese cities with the experience of cities elsewhere that are further along the motorization curve. Indications are that the series of changes in store will dramatically change urban life in China just as they have changed urban life in the cities of other countries. In some localities guidance of peripheral land development is receiving serious attention. At the national level, density standards for new development have been enacted. Nevertheless, the process of land development guidance needs to be continually reviewed and refocused in light of the current rapid change in the metropolitan areas of China.

REFERENCES

Construction Department of China. 2001. City Construction Statistics in China (2000). Beijing: Chinese Construction Industry Publishing Co.

Forman, R. T. T., et al. 2002. Road Ecology: Science and Solutions. Washington, D.C.: Island Press.

Gakenheimer, R., and D. DeLisi. 2000. Urban China Adjusting to Urbanization: A Place for Car Sharing. Cooperative Mobility Program, Massachusetts Institute for Technology, Cambridge.

Midgley, P. 1994. Urban Transport in Asia: An Operational Agenda for the 1990s. Technical Paper 224. World Bank, Washington, D.C.

National Research Council (NRC). 1997. Toward a Sustainable Transportation Future. Washington, D.C.: National Academy Press.

———. 1999. Transportation, Energy, and Environment: Policies to Promote Sustainability. Transportation Research Board. Washington, D.C.: National Academy Press.

Newman, P. W. G., and J. R. Kenworthy. 1989. Cities and Automobile Dependence: A Sourcebook. Aldershot, Hants, England: Gower Technical.

Ping, W. Y., and A. Murie. 1999. Housing Policy and Practice in China. London: Macmillan Press.

Rodrigue, J-P. 1994. The utility value of land use: Theoretical foundations and an application to Shanghai. Journal of Transport Geography 2 (1).

Shanghai Study Team. 2001. Research on the Development of Personal Use Car: Special Report—Shanghai Case Study. Shanghai, May.

State Statistical Bureau. 1991. China Statistical Yearbook 1991. Beijing: China Statistical Publishing House.

Technical Supervision Bureau and Construction Department of China. 1995. Code for Transport Planning on Urban Road. January 14.

Tiwari, G. 2002. Urban Transport for Growing Cities, High Capacity Bus Systems. Delhi: Macmillan.

World Bank. 2001. World Development Indicators 2001. Washington, D.C.: World Bank.

Wu, F., and A. G-O Yeh. 1999. Urban spatial structure in a transitional economy: The case of Guangzhou, China. Journal of the American Planning Association 65(autumn):377–394.

7

Environment and Health

One of the more obvious consequences of a rapidly growing vehicle fleet is its effect on the environment, particularly in cities. The air in most of China's large and medium-size cities is already unacceptably polluted, with the largest cities ranked among the most polluted in the world. As seen in other countries, this situation, when exacerbated by more vehicles on the road, will not improve unless governments take firm actions to control it. This chapter examines the atmospheric pollutants generated by vehicle operation.

EMISSIONS

The combustion of gasoline or diesel fuel in vehicle engines produces a variety of potentially harmful emissions. The amount and type of emissions depend on a variety of factors, including engine design, operating conditions, and fuel characteristics. Evaporative hydrocarbon emissions—from refueling, spills on heated engine parts, and so forth—can be just as harmful as those from the tailpipe.

Emissions from motor vehicles take two primary forms: (1) major gaseous and particulate air pollutants, which can be found in relatively high amounts in the atmosphere; and (2) air toxics, which usually are found in smaller amounts in the atmosphere but can have important effects on public health. The gaseous and particulate pollutants to which motor vehicles contribute include carbon monoxide (CO); ozone (O_3), through its atmospheric precursors volatile organic compounds (VOCs) and nitrogen oxides (NO_x); fine particulate matter PM_{10} and $PM_{2.5}$, particles smaller than

10 and 2.5 microns (μm) in aerodynamic diameter, respectively; and nitrogen dioxide (NO_2).[1] The air toxics emitted by motor vehicles include aldehydes (acetaldehyde, formaldehyde, and others), benzene, 1,3-butadiene, and a large number of substances known as polycyclic organic matter (including polycyclic aromatic hydrocarbons, or PAHs).

All of the pollutants emitted from motor vehicles also are produced by other sources such as industrial processes, electric power generation, and home heating. The contribution of motor vehicles to ambient levels varies, depending on the pollutant and the location. In most cases, motor vehicles are a major contributor (between 25 and 40 percent of the ambient levels), and for some pollutants—for example, carbon monoxide, ultrafine particles ($PM_{0.1}$), and 1,3-butadiene—motor vehicles tend to be the dominant source.

Although motor vehicles contribute a significant portion, if not the largest part, of most air pollutants, in certain circumstances they can contribute a substantially higher amount to personal exposure. In particular, in urban centers, along roadsides, and especially in urban street canyons in crowded central business districts, mobile sources can contribute 2 to 10 times as much as in general background situations.[2] For example, in England urban background levels of PM_{10} have been measured at 22–25 micrograms per cubic meter (μg/m^3), and street-side levels have been measured at 24–38 μg/m^3 (Department of the Environment, Transport, and Regions, 1999). Such a finding can have important implications for the potential acute health effects arising from exposure to these pollutants and for the chronic health effects on those people who spend a significant portion of their lives in these environments, especially the elderly, low-income, and other urban populations that may be especially sensitive to the effects of air pollution.

HEALTH EFFECTS

Research conducted over the past several decades has identified some of the effects that different pollutants have on human health, including those on the respiratory, neurological, and cardiac systems, and those that promote several types of cancer. One of the challenges

[1] Currently the ambient air quality standard is for nitrogen dioxide. Before 2000, however, there were standards for both nitrogen dioxide and nitrogen oxides, with different levels for each. A lot of the historical Chinese data are for NO_x.

[2] This is in general true, but for ozone, urban levels are generally lower than those found downwind of city centers, the result of the scavenging of the ambient ozone by high levels of ambient nitrogen oxides.

of understanding these effects is that they are usually experienced as part of a complex mixture of pollutants, and it is often difficult to disentangle the specific effects of one pollutant from the effects of other pollutants that follow similar spatial and atmospheric patterns (Health Effects Institute, 2000c). At the same time, it is apparent that not all members of the population are equally sensitive to such effects, and that some subgroups (e.g., the elderly, asthmatics, children, people with preexisting heart disease) may be at greater risk from exposure to air pollution than other adults.

Overall, the effects of these pollutants on an individual's health tend to be relatively small in comparison with other risk factors such as cigarette smoking. However, because a large number of people are exposed, the effects as a whole on overall public health are of sufficient magnitude to be of public concern. For example, one recent European analysis estimated that approximately 6 percent of mortality (40,000 deaths annually) in France, Austria, and Switzerland could be attributed to particulate air pollution alone, and about half of that could be attributed to exposure to vehicle emissions (Kunzli et al., 2000).

The pollutants of greatest concern from vehicles are carbon monoxide, hydrocarbons (HC), nitrogen oxides, ozone (which results from the emissions of hydrocarbons and nitrogen oxides), particles, and certain toxic hydrocarbons such as benzene.

Carbon Monoxide

Carbon monoxide (CO)—an odorless, invisible gas created when fuels containing carbon are burned incompletely—poses a serious threat to human health. Fetuses and anyone afflicted with heart disease are especially at risk. Because hemoglobin in the blood has an affinity for carbon monoxide that is 200 times greater than that for oxygen, carbon monoxide hinders the transport of oxygen from the blood into the tissues. Therefore, more blood must be pumped to deliver the same amount of oxygen. Numerous studies in humans and animals have demonstrated that people with weak hearts are subjected to additional strain by the presence of excess carbon monoxide in the blood. In particular, clinical health studies have shown that people suffering from angina pectoris and exposed to elevated levels of ambient carbon monoxide experience angina pain more quickly than usual.

Healthy people also are affected, but only at higher levels. Exposure to elevated CO levels is associated with impairment of visual perception, work capacity, manual dexterity, learning ability, and performance of complex tasks (U.S. EPA, 2000).

Nitrogen Oxides

As a class of compounds, the oxides of nitrogen (NO_x) are involved in a host of environmental interactions that have adverse effects on human health and welfare and the environment. Nitrogen dioxide (NO_2) has been linked with increased susceptibility to respiratory infection, increased airway resistance in asthmatics, and decreased pulmonary function (U.S. EPA, 1993a, 1995). Even short-term exposures to nitrogen dioxide have resulted in wide-ranging respiratory problems in schoolchildren—cough, runny nose, and sore throat are among the most common (Mostardi et al., 1981).

Nitrogen oxides also contribute to acid deposition, which can damage trees at high elevations and, by increasing the acidity of lakes and streams, severely damage aquatic life. Finally, NO_x emissions can contribute to increased levels of particulate matter by changing into nitric acid in the atmosphere and forming particulate nitrate.

Photochemical Oxidants (Ozone)

The science of ozone (O_3) formation, transport, and accumulation is complex. Ground-level ozone is produced and destroyed in a cyclical set of chemical reactions involving nitrogen oxides, VOCs, heat, and sunlight.[3] As a result, differences in NO_x and VOC emissions, their ratios, and weather patterns contribute to daily, seasonal, and yearly differences in ozone concentrations and differences from city to city. Many of the chemical reactions that are part of the ozone-forming cycle are sensitive to temperature and sunlight. When ambient temperatures and sunlight levels remain high for several days and the air is relatively stagnant, ozone and its precursors can build up and produce more ozone than typically would occur on a single high-temperature day.[4] Further complicating matters, ozone can be transported into an area from pollution sources hundreds of miles upwind, resulting in elevated ozone levels even in areas with low VOC or NO_x emissions.

VOCs are emitted from a variety of sources, including motor vehicles, chemical plants, refineries, factories, consumer and commercial products, and other industrial sources. VOCs also are emitted by natural sources

[3] Carbon monoxide also participates in the production of ozone, albeit at a much slower rate than most VOC and NO_x compounds.

[4] There is a growing concern that climate modification resulting from the increased buildup of greenhouse gases such as carbon dioxide may increase the amount of ozone produced from a given amount of nitrogen oxides and VOCs.

such as vegetation. Nitrogen oxides are emitted largely by motor vehicles, nonroad equipment, power plants, and other sources of combustion.

Based on a large number of studies, the U.S. Environmental Protection Agency (U.S. EPA) has identified several key health effects of human exposure to present levels of ozone (U.S. EPA, 1996a, 1996c). When inhaled, ozone can cause acute respiratory problems, including asthma attacks, significant temporary decreases in lung function of 15 to over 20 percent in some healthy adults, and inflammation of lung tissue, leading to increased hospital admissions and emergency room visits. Inhaling ozone also may impair the body's immune system defenses, making people more susceptible to respiratory illnesses. Children and outdoor workers are likely to be exposed to elevated ambient levels of ozone during exercise and therefore are at greater risk of experiencing adverse health effects.

In addition to its effects on human health, ozone is known to adversely affect the environment in many ways. These effects include reduced yields for commodity crops, fruits and vegetables, and commercial forests; deleterious effects on the ecosystem and vegetation in areas such as national parks; damage to urban grass, flowers, shrubs, and trees; reduced yields for tree seedlings and noncommercial forests; increased susceptibility of plants to pests; materials damage; and decreased visibility.

In addition to their contribution to ozone levels, emissions of certain hydrocarbons contain toxic air pollutants that may have a significant effect on public health.

Toxic Hydrocarbons

The U.S. Environmental Protection Agency recently reconfirmed that benzene is a known human carcinogen by all routes of exposure (U.S. EPA, 1998). Respiration is the major source of human exposure. Long-term respiratory exposure to high levels of ambient benzene concentrations has been shown to cause cancer of the tissues that form white blood cells. Leukemias, lymphomas, and other tumor types have been observed in experimental animals exposed to benzene by inhalation or oral administration. Exposure to benzene or its metabolites also has been linked to genetic changes in humans and animals (IARC, 1982) and increased proliferation of mouse bone marrow cells (Irons et al., 1992). The occurrence of certain chromosomal changes in persons with known exposure to benzene may serve as a marker for those at risk for contracting leukemia (Lumley et al., 1990).

U.S. EPA has classified formaldehyde as a probable human carcinogen based on limited evidence of carcinogenicity in humans and sufficient evidence of carcinogenicity in animal studies using rats, mice, hamsters, and monkeys (U.S. EPA, 1987). Epidemiological studies of occupationally exposed workers suggest that long-term inhalation of formaldehyde may be associated with tumors of the nasopharyngeal cavity (generally the area at the back of the mouth near the nose), nasal cavity, and sinuses. Research has demonstrated that formaldehyde produces mutagenic activity in cell cultures (U.S. EPA, 1993b).

The atmospheric chemistry of acetaldehyde is similar in many respects to that of formaldehyde (Ligocki and Whitten, 1991). Like formaldehyde, it is produced and destroyed by atmospheric chemical transformation. Acetaldehyde is classified by the U.S. EPA as a probable human carcinogen.

The pollutant 1,3-butadiene is formed in vehicle exhaust by the incomplete combustion of fuel. It was classified by the U.S. EPA as a Group B2 (probable human) carcinogen in 1985 (U.S. EPA, 1985). This classification was based on evidence from two species of rodents and epidemiological data.

Particulates

Particulate matter (PM) is a broad class of chemically and physically diverse substances that exist as discrete particles (liquid droplets or solids) over a wide range of sizes. Human-generated sources of particles include a variety that is either stationary or mobile. Particles may be emitted directly to the atmosphere or may be formed by transformations of gaseous emissions such as sulfur dioxide or nitrogen oxides. The major chemical and physical properties of particulate matter vary greatly with time, region, meteorology, and source category, thereby complicating any assessment of the health and welfare effects that might be related to various indicators of particulate pollution. At elevated concentrations, particulate matter can adversely affect human health, visibility, and materials. Components of particulate matter (e.g., sulfuric or nitric acid) contribute to acid deposition (U.S. EPA, 1996b).

The key health effects associated with particulate matter include premature death; aggravation of respiratory and cardiovascular disease, as indicated by increased hospital admissions and emergency room visits, school absences, lost work days, and restricted activity days; changes in lung function and increased respiratory symptoms; changes to lung tissues and structure; and altered respiratory defense mechanisms (U.S. EPA, 1996b). Most of these effects have been consistently associated with ambient PM concentrations, used as a measure of population exposure, in a large number of community epidemiological studies. Additional information and insights on these effects are provided by studies of animal toxicology and controlled human exposures to various constituents of particulate matter conducted at higher-than-ambient concentrations. Although the mechanisms by which particles produce effects are not well known, there is general agreement that the cardiorespiratory system is the major target of the effects of particulate matter.

People with infectious respiratory disease (e.g., pneumonia) are at greater risk of premature mortality and morbidity (e.g., hospitalization, aggravation of respiratory symptoms) from exposure to ambient particulate matter. Also, such exposure may increase a healthy person's susceptibility to respiratory infections. The elderly, children, and the asthmatic face even greater risks.

Fine and coarse fraction particles have fundamental physical and chemical differences. The fine fraction contains acid aerosols, sulfates, nitrates, transition metals, diesel exhaust particles, and ultrafine particles, and the coarse fraction typically contains high mineral concentrations, silica, and suspended dust. Exposure to coarse fraction particles is primarily associated with the aggravation of respiratory conditions such as asthma. Fine particles are most closely associated with health effects such as cardiopulmonary diseases.

The strongest evidence for ambient PM exposure health risks is derived from epidemiological studies (Health Effects Institute, 2000a, 2000b). Many have shown statistically significant associations of ambient PM levels with a variety of human health endpoints in sensitive populations, including mortality, hospital admissions and emergency room visits, respiratory illness and symptoms, and physiological changes in mechanical pulmonary function. The epidemiological science points to fine particulate matter being more strongly associated with acute conditions and premature mortality than coarse fraction particulate matter, which is associated with chronic health effects.

Time-series analyses strongly suggest a positive effect on daily mortality across the entire range of ambient PM levels. Relative risk estimates for daily mortality in relation to daily ambient PM concentration are consistently positive and statistically significant (at $PM_{0.05}$), across a variety of statistical modeling approaches and methods of adjustment for effects of relevant covariates such as season, weather, and copollutants.

Diesel Emissions

Diesel emissions deserve a special discussion because they tend to be a dominant source of mobile source cancer risk. In 1993 the U.S. Environmental Protection Agency determined a reference concentration to minimize the noncancer health effects of exposure to diesel exhaust. Based on information provided in the draft "Health Assessment Document for Diesel Emissions" and other sources of information, the U.S. EPA concluded that diesel particulate is a probable human carcinogen. The most compelling information to suggest a carcinogenic hazard is the consistent association observed between increased lung cancer and diesel exhaust exposure in certain workers laboring in the presence of diesel engines (Health Effects Institute, 1999). Approximately 30 individual epidemiological studies have shown increased lung cancer risks of 20–89 percent within the study populations. The analytical results of pooling the positive study results reveal that on average the lung cancer risks were increased by 33–47 percent. The magnitude of the pooled risk increase is not precise because of the uncertainties in the individual studies, the most important of which is a continuing concern about whether smoking effects were accounted for adequately. Although not all studies demonstrated an increased risk—6 of 34 epidemiological studies summarized by the Health Effects Institute (1995) reported relative risks of less than 1.0—the fact that an increased risk has been consistently noted in the majority of epidemiological studies strongly supports the determination that exposure to diesel exhaust is likely to pose a carcinogenic hazard to humans.

Additional evidence that supports treating diesel exhaust as a carcinogen at ambient levels of exposure is provided by the observation of small quantities of many mutagenic and some carcinogenic compounds in the diesel exhaust. A carcinogenic response to such agents is assumed not to have a threshold unless there is direct evidence to the contrary. In addition, there is evidence that at least some of the organic compounds associated with diesel particulate matter are extracted by lung fluids (i.e., are bioavailable) and therefore are dispensed in some quantity to the lungs and able to enter the bloodstream and travel to other sites in the body.

Several national and international agencies have designated diesel exhaust or diesel particulate matter as a "potential" (National Institute for Occupational Safety and Health) or "probable" (International Agency for Research on Cancer) human carcinogen (NIOSH, 1988, IARC, 1989). Based on the IARC findings, in 1990 the state of California identified diesel exhaust as a chemical known to the state to cause cancer, and after an extensive review in 1998 it listed diesel exhaust as a toxic air contaminant (California Environmental Protection Agency, 1998). The World Health Organization recommends that "urgent efforts should be made to reduce [diesel engine] emissions, specifically of particulates, by changing exhaust train techniques, engine design and fuel composition" (WHO, 1996). More recently, in its *Ninth National Toxicology Program Report on Carcinogens* the National Institute for Environmental Health Sciences added diesel particulate to its list of substances that are reasonably thought to be human carcinogens (NIEHS, 2001).

Another aspect of diesel particulate that is a cause for concern is its size. Approximately 80–95 percent of diesel particle mass is in the size range of

0.05–1.0 μm with a mean particle diameter of about 0.2 μm. These fine particles have a very large surface area per gram of mass, which make them excellent carriers for adsorbed inorganic and organic compounds that can effectively reach the lowest airways of the lung. Some 50–90 percent of the numbers of particles in diesel exhaust are in the ultrafine size range, from 0.005 to 0.05 μm, averaging about 0.02 μm. Ultrafine diesel particulate matter, which accounts for the majority of the number of particles, also accounts for 1–20 percent of the mass of diesel particulate matter.

CLIMATE CHANGE

Beyond direct adverse health effects, vehicle emissions are a source of other concerns. Among these is climate change, or the greenhouse effect. Greenhouse warming occurs when certain gases allow sunlight to penetrate to the Earth but partially trap the planet's radiated infrared heat in the atmosphere. Some such warming is natural and necessary. If there were no water vapor, carbon dioxide, methane, and other infrared absorbing (greenhouse) gases in the atmosphere trapping the Earth's radiant heat, the planet would be about 60 degrees Fahrenheit (or 33 degrees Celsius) colder, and life as we know it would not be possible. Naturally occurring greenhouse gases include water vapor, carbon dioxide (CO_2), methane (CH_4), nitrous oxide (N_2O), and ozone (O_3).

Several classes of halogenated substances that contain fluorine, chlorine, or bromine are also greenhouse gases, but they are, for the most part, solely a product of industrial activities. Chlorofluorocarbons (CFCs) and hydrochlorofluorocarbons (HCFCs) are halocarbons that contain chlorine, and halocarbons that contain bromine are known as halons. Other fluorine-containing halogenated substances include hydrofluorocarbons (HFCs), perfluorocarbons (PFCs), and sulfur hexafluoride (SF_6).

Although they do not have a direct global warming effect, several gases do influence the formation and destruction of ozone, which has a terrestrial radiation-absorbing effect. These gases include carbon monoxide, oxides of nitrogen, and nonmethane volatile organic compounds (NMVOCs).

Aerosols, extremely small particles or liquid droplets often produced by emissions of sulfur dioxide (SO_2), also can affect the absorptive characteristics of the atmosphere.

Although carbon dioxide, methane, and nitrous oxide occur naturally in the atmosphere, the atmospheric concentration of each has risen, largely as a result of human activities. Since 1800, atmospheric concentrations of these greenhouse gases have increased by 30, 145, and 15 percent, respectively (IPCC, 1996). This buildup has altered the composition of the Earth's atmosphere and may affect the global climate system.

Beginning in the 1950s, the use of CFCs and other ozone-depleting substances (ODSs) increased by nearly 10 percent a year, until the mid-1980s when international concern about ozone depletion led to the signing of the Montreal Protocol. Since then, the use of ODSs has declined rapidly, and they are being phased out completely. In contrast, the use of ODS substitutes, such as hydrofluorocarbons, perfluorocarbons, and sulfur hexafluoride, has grown significantly, and all have strong greenhouse forcing effects.

In late November 1995 Working Group 1 of the International Panel on Climate Change (IPCC) concluded that "the balance of evidence suggests that there is a discernible human influence on global climate" (IPCC, 1996).[5] In December 1997, acting on this consensus, countries around the world approved the Kyoto Protocol to the 1992 Climate Change Treaty. When, and if, the protocol is ratified by 55 nations, representing 55 percent of 1990 CO_2 emissions, 38 industrialized nations will be required to reduce, by between 2008 and 2012, their greenhouse gas emissions from the 1990 levels. The European Union would reduce emissions by 8 percent, the United States by 7 percent, and Japan by 6 percent.[6] Some nations would face smaller reductions, and a few would not face any for the moment. As a group, the industrialized nations would cut back on the emissions of such gases by just more than 5 percent. Emissions of six gases would be affected: carbon dioxide, methane, nitrous oxide, and three halocarbons used as substitutes for ozone-damaging chlorofluorocarbons.

The greenhouse gases most closely identified with the transportation sector include carbon dioxide, nitrous oxide, and methane (see Table 7-1 for the global warming potential of nitrous oxide and methane relative to carbon dioxide). Other vehicle-related pollutants also contribute to global warming, but their quantification has been more difficult. These include carbon monoxide, nonmethane hydrocarbons (NMHC), and nitrogen dioxide. In the original (1990) IPCC report, global warming potentials (GWPs) were attributed to these gases (Shine et al., 1990). Because of difficulty reaching agreement on the appropriate quantification, specific GWPs for these gases were not contained in the most recent IPCC report (Table 7-1).

In most countries, over 90 percent of the global warming potential of the direct-acting greenhouse gases from the transportation sector comes from carbon dioxide, and therefore the global warming potential from

[5] In its most recent draft report, the IPCC has removed the qualifier to say, "There is a discernible human influence on global climate."

[6] The U.S. government recently indicated that it will not ratify the Kyoto Protocol but that it intends to suggest an alternative approach.

TABLE 7-1 IPCC's Global Warming Potential (GWP) for Carbon Monoxide, Methane, Nonmethane Hydrocarbons, Nitrogen Dioxide, and Nitrous Oxide

GWP	Carbon Monoxide (CO)	Methane (CH_4)	Nonmethane Hydrocarbons (NMHC)	Nitrogen Dioxide (NO_2)	Nitrous Oxide (N_2O)
20-year horizon	7	56	31	30	280
100-year horizon	3	21	11	7	310
500-year horizon	2	6.5	6	2	170

NOTE: The time horizon is the time period over which the GWP is measured relative to carbon dioxide. Different gases have different lifetimes in the atmosphere.
SOURCE: IPCC (1996).

transportation is most closely related to fuel economy. The transportation sector is responsible for about 17 percent of global CO_2 emissions, and these emissions are increasing in virtually every part of the world.

Even the potential global warming benefits of diesel vehicles, because they are more fuel-efficient than gasoline-fueled vehicles, have been undercut by recent studies, which indicate that diesel particles may, by reducing cloud cover and rainfall, more than offset any CO_2 advantage they offer. As James Hansen and his colleagues at the U.S. National Aeronautics and Space Administration (NASA) have noted, "Black carbon[7] reduces aerosol albedo, causes a semi-direct reduction of cloud cover, and reduces cloud particle albedo" (Hansen et al., 2001). Tight control of diesel particulate emissions would reduce their negative greenhouse effect and allow full greenhouse benefits from the CO_2 advantage that diesels provide.

AIR QUALITY

One result of the rapid growth of China's vehicle fleet has been a significant increase in urban air pollution. In spite of significant advances in industrial pollution control, air pollution in the major Chinese cities remains a serious problem and in some cases may actually be worsening. It is generally characterized as a shift from coal-based pollution to vehicle-based pollution.

Based on the available data, it is clear that the national NO_x air quality standards are currently exceeded across large areas in China, including but not limited to high traffic ones. Before 1992, the annual average NO_x

[7] Black carbon pollution is the release of particulates from burning fuel into the air.

TABLE 7-2 Ozone Concentration in Beijing, 1997–1999

	Number of Nonattainment Days	Number of Nonattainment Hours	Maximum Hourly Concentration (μg/m^3)
1997	71	434	346
1998	101	504	384
1999	119	777	—

NOTE: "Nonattainment" hours means time that the respective air quality standard was exceeded in Beijing, which occurred on the indicated number of distinct nonattainment days. The maximum hourly concentration was the highest value observed. — = not available; (μg/m^3) = micrograms per cubic meter.
SOURCE: He Kebin, Tsinghua University, Beijing.

concentration in Shanghai was lower than 50 micrograms per cubic meter (μg/m^3), which complied with the Chinese Class II air quality standard. But since 1995 the NO_x concentration has been gradually increasing, from 51 μg/m^3 in 1995 to 59 μg/m^3 in 1997 (Shanghai Municipal Government, 1999).

In Beijing, NO_x concentrations within the Second Ring Road that encircles the city center increased from 99 μg/m^3 in 1986 to 205 μg/m^3 in 1997, more than doubling in a decade. Moreover, CO and NO_x concentrations on the urban trunk traffic roads and interchanges exceed national environmental quality standards year-round (Beijing Municipal Environmental Protection Bureau et al., 1999). National air quality standards for particulates also are frequently exceeded—primarily because of coal and charcoal burning.

Recent data indicate as well that standards for ozone, formed by the photochemical reaction of nitrogen oxides and hydrocarbons, have been exceeded in several metropolitan areas during the last decade (see Table 7-2, which shows a clear upward trend in Beijing).

On average, mobile sources are currently contributing 45–60 percent of NO_x emissions and about 85 percent of CO emissions in typical Chinese cities (Project of China Environment Technological Assistance Loaned by the World Bank B-9-3, 1997). For example, data collected in Shanghai show that in 1996, of the total air pollution load in the downtown area, vehicles emitted 86 percent of the carbon monoxide, 56 percent of the nitrogen oxides, and 96 percent of the nonmethane hydrocarbons (Shanghai Municipal Government, 1999). In Beijing in recent years, the NO_x concentration shows a clear upward trend. In 1997 the annual average NO_x concentration was 133 μg/m^3, the average concentration during the heating season was 191 μg/m^3, and that during the non-heating season was 99 μg/m^3. These emissions were, respectively, 73 percent, 66 percent, and 80 percent higher than those 10 years ago. The annual daily average NO_x concentration in 1998 was 14.3 percent higher than in 1997. Because the amount of coal burning has remained stable for many years, Beijing local authorities attribute the increases to vehicular emissions (Beijing Municipal Environmental Protection Bureau et al., 1999). According to the Beijing Municipal Environmental Protection Bureau, "In 2000, NO_x emissions by motor vehicles accounted for 43% of the total and CO emissions, 83%. As the vehicle discharges pollutants at low altitude, it contributes to 73% and 84% of the effect on environmental quality."[8]

In response to the air pollution problem, China has initiated a motor vehicle pollution control effort. It has moved quickly to eliminate the use of leaded gasoline and recently introduced European Emission Standard I (Euro I) for new cars and trucks. It will introduce the Euro II standards in 2004.[9] Nevertheless, the emissions requirements for new vehicles lag behind those of the industrialized world by about a decade. Furthermore, without additional improvements in fuel quality, greater tightening of new vehicle standards will be difficult (see Chapter 5). Another factor is that in China road conditions and maintenance practices are exacerbating the air pollution problem.

The Chinese government has expended a great effort to mitigate the primary pollutants such as SO_2, NO_x, and PM_{10} in many cities—the levels of these pollutants are measured routinely by the central and local governments. However, secondary pollutants such as photochemical smog (ozone) and the fine particles emitted by primary and secondary sources are far greater threats to human health. Vehicular emissions contribute significantly to the formation of ground-level ozone and fine particles, as well as to an increase in greenhouse gases.

IMPLICATIONS OF CHINA'S VEHICLE GROWTH FOR FUTURE EMISSIONS AND FUEL CONSUMPTION

As indicated in Chapter 2, China is anticipating a threefold to sevenfold increase in its vehicle fleet, not including motorcycles, between 2002 and 2020 (see Table 2-1). The number of cars, in particular, is expected to increase by three to nine times in the same time period. This section summarizes emissions and fuel consumption estimates that are based on the vehicle characteristics in the five-year plan for the automotive industry. Using the medium-growth scenario from Chapter 2, Figure 7-1 indicates

[8] Yu Xiaoxuan, Beijing Municipal Environmental Protection Bureau.

[9] Beijing will introduce Euro II standards one year earlier than the rest of the country in 2003.

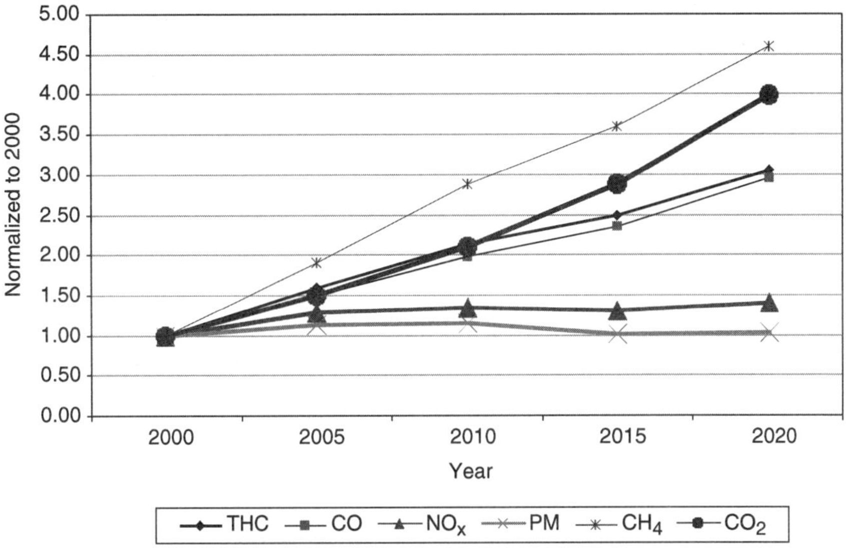

FIGURE 7-1 Motor vehicle emissions in China, 2000–2020. NOTE: THC = total hydrocarbons; CO = carbon monoxide; NO_x = nitrogen oxides; PM = particulate matter; CH_4 = methane; CO_2 = carbon dioxide. SOURCE: Calculations by Michael P. Walsh.

that motor vehicle emissions of carbon dioxide will about quadruple between 2000 and 2020; CO and hydrocarbon levels will about triple; and NO_x and PM levels will stay essentially at the currently high levels.

For light-duty vehicles (and the medium-growth scenario), the pollution trends are somewhat better (see Table 7-3). However, CO and NO_x levels will increase, and CO_2 emissions are estimated to be more than three and a half times higher in 2020 than in 2000. If the highest-growth scenario should become reality, motor vehicle emissions of all pollutants would increase and carbon dioxide would skyrocket (Figure 7-2).

The potential for reducing fuel consumption and CO_2 emissions was investigated using two cases that focused on light-duty vehicles. In case 1, it was assumed that starting in 2005 the fuel economy of all new gasoline-fueled cars and light trucks improved by 2 percent a year. Case 2 further assumed that starting in model year 2010, 5 percent of highly efficient cars and light trucks (and increasing by 5 percent a year) achieve fuel consumption of 80 miles per gallon (mpg). As illustrated in Figure 7-3, under these scenarios the growth in fuel consumption falls but several more years will be needed for the full effects to be felt. For case 2, by 2020 from 12 billion to over 30 billion gallons of fuel will be saved compared with the base case, depending on which vehicle growth rate occurs.

TABLE 7-3 Light-duty Vehicle Emissions Trends in China, 2000–2020

	2000	2005	2010	2015	2020
THC	1.00	1.01	0.95	0.83	0.85
CO	1.00	1.13	1.22	1.29	1.56
NO_x	1.00	1.19	1.31	1.30	1.42
PM	1.00	1.01	1.04	0.94	0.96
CO_2	1.00	1.41	1.97	2.77	3.87

NOTE: Values shown are emissions normalized to base year 2000. THC = total hydrocarbons; CO = carbon monoxide; NO_x = nitrogen oxides; PM = particulate matter; CO_2 = carbon dioxide.

SOURCE: Calculations by Michael P. Walsh using mid-range scenario of Chapter 2, assuming 8 percent growth of GDP.

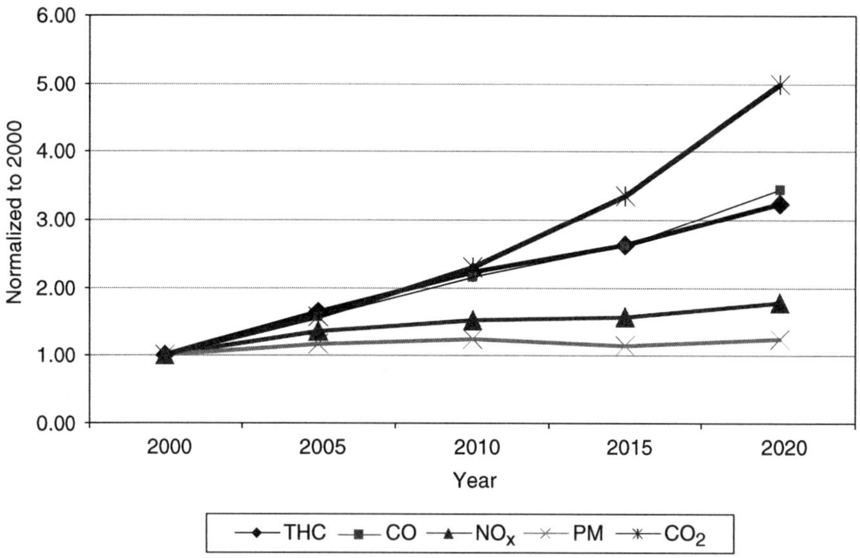

FIGURE 7-2 Motor vehicle emissions in China—European standards in 2010, light-duty fuel economy improvements starting in 2005. NOTE: THC = total hydrocarbons; CO = carbon monoxide; NO_x = nitrogen oxides; PM = particulate matter; CO_2 = carbon dioxide. SOURCE: Calculations by Michael P. Walsh.

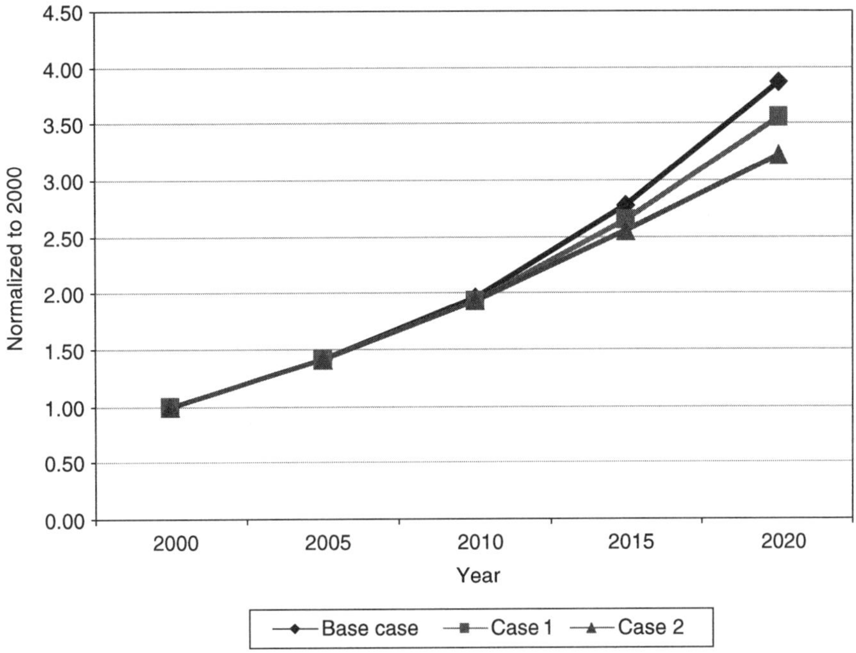

FIGURE 7-3 Light-duty carbon dioxide emissions, alternative scenarios. Base case: year 2000 trends continue; case 1: fuel economy of fleet improves at rate of 2 percent per year; case 2: same as case 1, but with the addition to the fleet of cars capable of 80 miles per gallon at a rate of 5 percent of new cars per year. SOURCE: Calculations by Michael P. Walsh

In conclusion, the various pollutants emitted by vehicles are a large and potentially growing source of air pollution in China. They already account for a substantial fraction of emissions contributing to excessively high ambient levels of air pollution. Investigators have shown that these pollutants have a measurable negative effect on the public health. Even with the currently adopted emissions standards, Euro II by 2004, and a 10 percent improvement in vehicle fuel economy, as called for in the five-year plan, emissions of all pollutants will increase if high growth occurs. Even if growth is constrained to the medium case, all pollutants but particulate matter are expected to increase, although if the light-duty diesel fleet grows significantly, particulates and nitrogen oxides will increase more and CO_2 emissions will be lower. As a result, efforts to reduce vehicle emissions must continue, to avoid the concomitant impacts on public health and the environment. If growth can be constrained to the medium case and if emissions standards are aligned with those of the European Union by 2010, it should be possible to actually reduce vehicle emissions of total hydrocarbons, carbon monoxide, nitrogen oxides, and particulate matter. A complementary vehicle fuel efficiency program will be needed to slow down the growth in CO_2 emissions and fuel consumption.

According to a recent study by Shao et al. (2001), the only way for Guangzhou City to achieve its air quality targets by 2010 is to advance the implementation of Euro III standards to as early as 2004. Acceleration of the implementation schedule was found to be technically feasible, because the vehicle technologies needed to meet Euro III are already available. Such a step also would advance China's prospects of meeting Euro IV standards by 2010. However, considerations of fuel quality, infrastructure, and economic cost must be addressed.

In view of the very rapid growth in the vehicle fleet forecast for the next two decades, China's environment could face severe strains and significant public health consequences unless vehicle technology is substantially upgraded and fuel quality improved. Similarly, fuel consumption and greenhouse gas emissions will increase dramatically without substantial improvements in vehicle technology. China should strongly consider developing the appropriate mix of performance standards and incentives necessary to leapfrog from today's modest requirements to the global state of the art as rapidly as possible.

REFERENCES

Beijing Municipal Environmental Protection Bureau, Beijing Municipal Public Security and Traffic Administration Bureau, and Beijing Urban Planning, Design and Research Academy. 1999. Urban Transport and Environment in Beijing. January 15.

California Environmental Protection Agency. 1998. Proposed Identification of Diesel Exhaust as a Toxic Air Contaminant. Appendix III, Part A: Exposure Assessment. California Environmental Protection Agency, California Air Resources Board, April 22.

Department of the Environment, Transport, and Regions. 1999. Source Apportionment of Airborne Particulate Matter in the United Kingdom. Report of the Airborne Particles Expert Group. London, January.

Hansen, J., M. Sato, R. Ruedy, A. Lacis, and V. Oinas. 2001. Global Warming in the 21st Century: An Alternative Scenario. Online. NASA Goddard Institute for Space Studies. Available at www.giss.nasa.gov/research/impacts/altscenario/. Accessed August 30, 2002.

Health Effects Institute. 1995. Diesel Exhaust: A Critical Analysis of Emissions, Exposure, and Health Effects. Cambridge, Mass., April.

———. 1999. Diesel Emissions and Lung Cancer: Epidemiology and Quantitative Risk Assessment. A Special Report of the Institute's Diesel Epidemiology Expert Panel. Cambridge, Mass., June.

———. 2000a. The National Morbidity, Mortality, and Air Pollution Study. Research Report 94, Part II. Cambridge, Mass.

———. 2000b. Reanalysis of the Harvard Six Cities and American Cancer Society Studies. A Special Report of the Institute's Particle Epidemiology Reanalysis Project. Cambridge, Mass.

———. 2000c. Strategic Plan for the Health Effects of Air Pollution, 2000–2005. Cambridge, Mass.

International Agency for Research on Cancer (IARC). 1982. Some Industrial Chemicals and Dyestuffs. Vol. 29, IARC Monographs on the Evaluation of Carcinogenic Risk of Chemicals to Humans. IARC, World Health Organization, Lyon, France.

———. 1989. Diesel and Gasoline Engine Exhausts and Some Nitroarenes. Vol. 46, Monographs on the Evaluation of Carcinogenic Risks to Humans. IARC, World Heath Organization, Lyon, France.

International Panel on Climate Change (IPCC). 1996. Climate Change 1995: The Science of Climate Change, J. T. Houghton et al., eds. New York: Cambridge University Press.

Irons, R. D., W. S. Stillman, D. B. Colagiovanni, and V. A. Henry. 1992. Synergistic action of the benzene metabolite hydroquinone on myelopoietic stimulating activity of granulocyte/macrophage colony-stimulating factor *in vitro.* Proceedings of the National Academy of Sciences 89:3691–3695.

Kunzli, N., R. Kaiser, S. Medina, M. Studnicka, O. Chanel, P. Filliger, M. Herry, F. Horak Jr., V. Puybonnieux-Texier, P. Quénel, J. Schneider, R. Seethaler, J-C. Vergnaud, and H. Sommer. 2000. Public-health impact of outdoor and traffic-related air pollution: A European assessment. Lancet 356 (September 2).

Ligocki, M. P., and G. Z. Whitten. 1991. Atmospheric Transformation of Air Toxics: Acetaldehyde and Polycyclic Organic Matter. SYSAPP-91/113. Systems Applications International, San Rafael, Calif.

Lumley, M., H. Barker, and J. A. Murray. 1990. Benzene in petrol. Lancet 336:1318–1319.

Mostardi, R. A., et al. 1981. The University of Akron study on air pollution and human health effects. Archives of Environmental Health (September/October).

National Institute for Environmental Health Sciences (NIEHS). 2001. Ninth National Toxicology Program Report on Carcinogens. Online. Available at 157.98.13.104/roc/ninth/rahc/diesel.pdf. Accessed August 29, 2002.

National Institute for Occupational Safety and Health (NIOSH). 1988. Carcinogenic effects of exposure to diesel exhaust. NIOSH Current Intelligence Bulletin 50. DHHS (NIOSH) Publication No. 88-116. Centers for Disease Control, Atlanta.

Project of China Environment Technological Assistance Loaned by the World Bank B-9-3. 1997. Strategy Analysis of Vehicular Emission Pollution Control in China (Summary Report). Environmental Engineering Department, Tsinghua University, Beijing, October.

Shanghai Municipal Government. 1999. Strategy for Sustainable Development of Urban Transportation and Environment—for a Metropolis with Coordinating Development of Transportation and Environment toward the 21st Century.

Shao M., Zhang Y., and R. Raufer. 2001. Control strategies for NO_x emissions in Guangzhou, China. Natural Resources Forum 25(2):157–165.

Shine, K. P., R. G. Derwent, D. J. Wuebbles, and J. J. Morcrette. 1990. Radiative forcing of climate. In Climate Change, the IPCC Scientific Assessment, J. T. Houghton, G. J. Jenkins, and J. J. Ephraums, eds. Cambridge: Cambridge University Press.

U.S. Environmental Protection Agency (U.S. EPA). 1985. Mutagenicity and carcinogenicity assessment of 1,3-butadiene. EPA Report No. EPA/600/8-85/004F. Office of Health and Environmental Assessment, EPA, Washington, D.C.

———. 1987. Assessment of Health Risks to Garment Workers and Certain Home Residents from Exposure to Formaldehyde. Office of Pesticides and Toxic Substances, EPA, Washington, D.C., April.

———. 1993a. Air Quality Criteria for Oxides of Nitrogen. EPA Report No. EPA/600/8-91/049aF. Washington, D.C.

———. 1993b. Motor Vehicle-Related Air Toxics Study. EPA Report No. EPA 420-R-93-005. Office of Mobile Sources, EPA, Ann Arbor, Mich., April.

———. 1995. Review of National Ambient Air Quality Standards for Nitrogen Dioxide, Assessment of Scientific and Technical Information. OAQPS Staff Paper. EPA-452/R-95-005. Office of Air Quality Planning and Standards, EPA, Washington, D.C.

———. 1996a. Air Quality Criteria for Ozone and Related Photochemical Oxidants. EPA Report No. EPA/600/P-93/004aF. Washington, D.C.

———. 1996b. Air Quality Criteria for Particulate Matter. EPA Report No. EPA/600/P-95/001aF. Washington, D.C.

———. 1996c. Review of National Ambient Air Quality Standards for Ozone, Assessment of Scientific and Technical Information. OAQPS Staff Paper. EPA-452/R-96-007. Office of Air Quality Planning and Standards, EPA,Washington, D.C.

———. 1998. Carcinogenic Effects of Benzene: An Update. EPA Report No. EPA/600/P-97/001F. National Center for Environmental Assessment, EPA, Washington, D.C.

———. 2000. Air Quality Criteria for Carbon Monoxide. Office of Research and Development, EPA, Washington, D.C., June.

World Health Organization (WHO). 1996. Diesel Fuel and Exhaust Emissions: International Program on Chemical Safety. Geneva, Switzerland.

8

The Role of Government

In every country, transport is largely in the public sector. Even vehicle manufacturing and use, the part that is commonly dominated by the private sector, is highly dependent on government services and government-provided infrastructure and is subject to government regulation and monitoring.

In China, the relationship between the government and automotive industry is in transition. Increasing uncertainty about the future structure of the industry and the growing dependence on market forces may test the relationships between industry and government and among different levels of government. Who will have responsibility for the dealing with the consequences of rapid motorization has not been clearly established. Meanwhile, cities are decentralizing, land markets are changing, and the dynamical forces involved are not well understood. The institutions dealing with motorization have limited experience with land markets, accelerating migration, and other externalities. The government will have to deal with conflicts between personal desires and the public good, between expanded vehicle ownership and equitable land use management.

STRATEGIC FRAMEWORK

Governments play a significant role in shaping the development of their domestic industries, and in determining the standards governing individual products and the impact of those products on the environment. For the automotive industry, governments around the world have used the following tools, singly or in combination:

- tariff and nontariff import barriers—used by many countries, including China, to protect their domestic industry during the early stages of development
- vehicle and fuel taxes—used to support or discourage the purchase and use of vehicles and fuels and to favor one technology or fuel over another
- prescriptive and performance standards—used by government to force certain vehicle attributes (e.g., low emissions, good fuel economy) or technologies (e.g., air bags)
- direct and indirect investment—used to assist industry with capital, tax relief, and support for research and development (e.g., the U.S. Partnership for a New Generation of Vehicles program in which government joined industry in funding research) or in building assembly plants.

One area in which most governments have intervened is reducing air pollution. Generally, a motor vehicle pollution control program seeks to reduce vehicle emissions to the degree necessary to achieve healthy air quality as rapidly as possible within the practical limits of effective technological, economic, and social constraints. A comprehensive strategy to achieve this goal usually includes four key components: increasingly stringent emissions standards for new vehicles, which require new technology; specifications for clean fuels; programs to ensure proper maintenance of in-use vehicles; and traffic and demand management (see Figure 8-1 and Box 8-1).

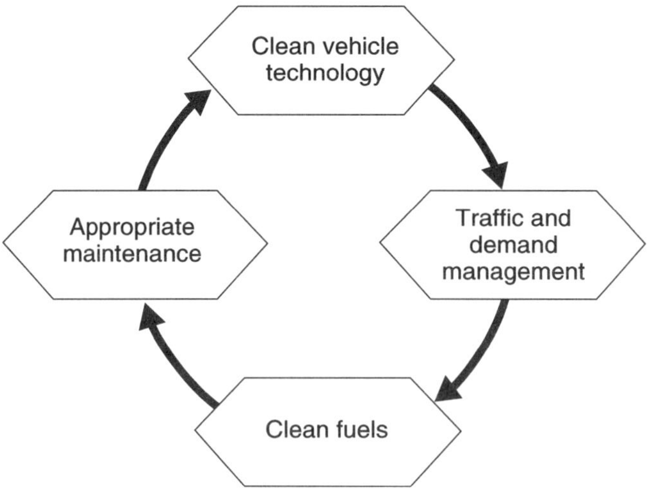

FIGURE 8-1 Elements of a comprehensive vehicle pollution control strategy. SOURCE: Michael P. Walsh.

Because vehicles are, by nature, mobile and therefore capable of being driven from one area to another, and because a proliferation of standards would be very costly to the manufacturer and to the purchaser of the product, new vehicle emissions standards are usually set by national governments to apply to a country as a whole (the state of California being a notable exception in the United States) or even to a group of countries. For similar reasons, clean fuels are mandated at the national level. The other two components, vehicle maintenance and traffic management, are usually the responsibilities of local governments and are applied in ways to respond to local air pollution problems. These can include measures such as:

- restrictions on the use of vehicles, including both cars and trucks
- high-occupancy vehicle lanes to encourage ride sharing
- inspection and maintenance programs, including protocols and training
- mandatory or voluntary retrofits of pollution control technology, with appropriate training
- recycling programs
- parking restrictions and parking taxes.

Details of the vehicle emissions control programs in the United States, Europe, and Japan are summarized in the next section. That summary is followed by a review of these countries' programs to reduce fuel consumption and carbon dioxide (CO_2) emissions. The chapter concludes with a brief look at inspection and maintenance programs, which is expanded in the appendix to this chapter, and, finally, a review of industry-government partnerships in the United States and Europe.

SUMMARY OF WORLDWIDE GOVERNMENT EMISSIONS STANDARDS

Because the United States, European Union (EU), and Japan each base their emissions regulations on different test procedures based on presumed typical driving patterns, it is difficult to compare precisely the stringency of those regulations. Tables 8-1 through 8-3 summarize the passenger car and heavy truck requirements in each region for nitrogen oxides (NO_x) and particulate matter (PM).

Table 8-1 indicates that the NO_x standards for passenger cars will become quite low in all three regions by 2005. However, the number of kilometers over which the European and Japanese governments require that vehicles in use meet emissions standards appear to be substantially lower

BOX 8-1
Transportation Planning and Traffic Management

Governments play an important role in effecting changes in land use that can reduce vehicle miles traveled (VMT). Over the long term, regulators should promote effective town and city land use planning that integrates mass transit options, promotes safe passage for pedestrians and cyclists, and uses design features to minimize single-occupancy vehicle use. The challenge is to make modes of living and working that reduce daily travel needs attractive to the public.

The many possible options for reducing VMT in urban areas include the following:

- *Internalizing costs.* Internalizing the cost of driving means shifting the expense to the consumer. Such a shift can be achieved through congestion pricing at automated tolls and parking facilities, reduced free parking and increased parking rates in general. "Pay-as-you-drive" insurance systems also are emerging as effective ways to distribute driving costs fairly. One example of these costs is parking. In the United States, roughly 90 percent of employee parking is either subsidized by the employer or free. Many areas have begun to increase the cost of parking to drivers, thereby internalizing the costs. Alternatively, some employers have implemented "cash out" policies, providing employees with the cash value of parking. Some other externalized costs of driving are the costs of policing and motorist protection; uninsured accidents; noise; vibration damage to structures; pollution damage to human health, crops, and structures; and petroleum industry subsidies.
- *Traffic management.* Traffic management strategies that increase vehicle occupancy directly reduce VMT, congestion, vehicle operating time, and therefore emissions. The special travel lanes for high-occupancy vehicles established in many urban areas to encourage car pooling have reduced the number of cars operating during peak travel periods. Advanced sensing systems at traffic lights facilitate the optimum utilization of urban street and intersection capacity, reducing congestion and the associated idling time. Many cities also have expanded the size of cyclist and pedestrian rights of way, while reducing street width and intersection size.
- *Increasing public transit use.* Public transportation provides an alternate means of mobility. The use of public transportation can be enhanced in three ways: by expanding public transit systems or making them more user-friendly; by increasing public awareness or acceptance of public transit; and by subsidizing the cost of consumers' use of public transit. Public awareness can be raised through traditional advertising campaigns, publicity events, and "free-ride" days that familiarize the public with transit systems. Subsidization of public transit commuting costs, either directly or through company programs, also has been effective in increasing transit

use levels. Often, enhancements to transit systems are most effective when coupled with other mechanisms, such as reduced parking availability, higher parking costs, or higher tolls. Promotion of car and van pooling also increases vehicle occupancy and reduces VMT. Government agencies, private corporations, and other institutions have offered this strategy.

- *Altering land use patterns.* Research clearly shows that gasoline use increases as decentralization increases, simply because people have farther to travel. A survey of 32 major cities around the world found that the residents of American cities consume nearly twice as much gasoline per capita as Australians (Kenworthy and Laube, 1999), nearly four times as much as residents of the more compact European cities, and 10 times that of those living in very compact Asian cities. Moreover, average living area per person was closely correlated with average gasoline consumption per capita. Land use patterns can be changed in ways that significantly reduce VMT and improve quality of life by providing for nearby amenities, such as pedestrian malls, and the preservation of undeveloped land. There are many ways to approach this task, including limiting development on urban fringes; offering location-specific mortgages that reward homeowners for buying in certain areas; creating more densely populated, mixed-use neighborhoods; changing vehicle rights of way to pedestrian malls; and reducing the number of parking spaces. Telecommuting also is a form of changing land use, in that it moves the office not just closer to the home, but into the home itself.

than the actual number of kilometers accumulated during average vehicle lifetimes. A key element of the low emissions level requirements has been the shift to very low-sulfur gasoline. Japan has traditionally had very low levels, usually below 30 parts per million (ppm) sulfur, but the EU recently capped levels at 50 ppm by 2005 and the United States capped them at 80 ppm with a 30 ppm average. The EU also is in the late stages of a process that will likely cap sulfur levels for both gasoline and diesel fuel at 10 ppm.

As for diesel-powered passenger cars, it is clear that the European Union and Japan, while substantially tightening their requirements over the next several years, will maintain substantially weaker NO_x requirements for diesel than for gasoline-fueled vehicles, unlike the United States (see Table 8-1 and Box 8-2). Similarly, the diesel car particulate requirements in the European Union and Japan are more lenient than those in the United States (Table 8-1). It appears that in Europe only the heavier diesel cars will require PM filters. In Japan a new round of standards will be introduced in 2005 to address particulates.

TABLE 8-1 Passenger Car Emissions Standards, Nitrogen Oxides (Gasoline and Diesel) and Particulate Matter (Diesel), United States, European Union, and Japan

	Standard	Year of Introduction	NO_x, Gasoline (g/km)	NO_x, Diesel (g/km)	PM, Diesel (g/km)	Vehicle Useful Life (thousand km)
U.S.— national	Tier 1	1994	0.373	0.777	0.062	160
	NLEV	2001	0.186	0.777	0.050	160
	Tier 2	2004	0.043	0.043	0.006	193
U.S.—California	TLEV	1994	0.373	0.373	0.050	160
	LEV	1994	0.186	0.186	0.050	160
	ULEV	1994	0.186	0.186	0.025	160
	LEV2	2004	0.043	0.043	0.006	193
	ULEV2	2004	0.043	0.043	0.006	193
	SULEV	2004	0.012	0.012	0.006	193
European Union	Euro III	2000	0.150	0.500	0.050	80
	Euro IV	2005	0.080	0.250	0.025	100
Japan	Japan	2002	0.080	0.280	0.052	80
		2005	0.050	0.140	0.013	80

NOTE: NO_x = nitrogen oxides; PM = particulate matter; g/km = grams per kilometer; NLEV = national low-emission vehicle; TLEV = transitional low-emission vehicle; LEV = low-emission vehicle; ULEV = ultra-low-emission vehicle; SULEV = super ultra-low-emission vehicle.

SOURCE: Daisho Yasuhiro, Waseda University.

BOX 8-2
Diesel Cars

The popularity of diesel cars varies widely worldwide, largely depending on government policies. They are found most commonly in Europe, accounting for one-third of car sales (and over half in some countries). In Japan they account for about 10 percent of sales, and in United States for only about 2 percent of light-duty vehicle sales (almost all light trucks).

A comparison of diesel cars and gasoline-powered cars reveals the advantages and disadvantages of diesel. In the area of pollution, diesel engines have inherently lower carbon monoxide and hydrocarbon emissions, and inherently higher particulate matter (PM) and nitrogen oxide (NO_x) emissions. In fact, Europe and Japan have adopted emissions standards that specifically allow diesel engines to emit higher levels of nitrogen oxides and particulate matter. Thanks to new technology, PM emissions from diesel engines have begun to fall dramatically, so that in future years if particulate filters are applied across the board, diesel emissions will be similar to those from gasoline engines,[1] but NO_x emissions from diesel engines are expected to remain higher than those from gasoline engines. Because diesel engines are inherently more energy efficient, they have lower carbon dioxide emissions than gasoline engines. In other performance characteristics, diesel engines are now roughly comparable to gasoline engines, including noise of operation.

Diesel engines are somewhat more costly to manufacture than gasoline engines, but the higher costs are countered by lower fuel costs. In regions where diesel fuel prices are significantly lower than gasoline prices (a result of political and economic tax policies but not fundamental cost differences), or where vehicles are used intensively, diesel cars will have strong cost advantages. In China, diesels are likely to proliferate if diesel fuel is priced lower than gasoline (which is the case in most countries but not the United States), if stringent fuel economy standards are adopted, and if more lenient NO_x standards are adopted for diesel. Otherwise, diesel cars are likely to be scarce in China.

[1] The currently adopted Euro IV standards, which go into effect for cars in 2005, are not stringent enough to require filters on all but the largest diesels.

Great progress is being made in reducing heavy-duty vehicle diesel emissions (Tables 8-2 and 8-3). Clearly, the major countries of the world have concluded that fundamental advances in heavy truck pollution controls and post–combustion technology are both necessary and feasible. A critical precondition of emission reduction, however, will be the introduction of very low or near-zero sulfur levels in fuel.

TABLE 8-2 Heavy-duty Diesel Nitrogen Oxide (NO_x) Standards, United States, European Union, Japan, 1990–2010 (grams per kilowatt-hour)

Model Year	United States	European Union[a]	Japan
1990	8.1	15.8	n.a.
1991	6.7	15.8	n.a.
1992	6.7	15.8	n.a.
1993	6.7	9	n.a.
1994	6.7	9	6
1995	6.7	9	6
1996	6.7	7	6
1997	6.7	7	6
1998	5.4	7	4.5
1999	5.4	7	4.5
2000	5.4	5	4.5
2001	5.4	5	4.5
2002	5.4	5	4.5
2003	2.7	5	3.38
2004	2.7	5	3.38
2005	2.7	3.5	2.0
2006	2.7	3.5	2.0
2007	0.27	3.5	2.0
2008	0.27	2	2.0
2009	0.27	2	2.0
2010	0.27	2	2.0

[a] Euro IV from 2005 and Euro V from 2008.

NOTE: n.a. = not applicable.

SOURCE: Daisho Yasuhiro, Waseda University.

United States

As indicated in Table 8-1, a notable exception to the rule that standards are normally set for an entire country is the provision in the U.S. Clean Air Act that allows the state of California to adopt its own vehicle emissions regulations. California is considered a unique case for several reasons. First, California has consistently had the most serious motor vehicle-related air pollution problems in the United States.[1] Second, Califor-

[1] Even today, despite the most aggressive car pollution control program in the world and despite significant progress, California's air pollution problems remain serious, and Los Angeles is consistently ranked as one of the most polluted cities in the United States.

TABLE 8-3 Heavy-duty Diesel Particulate Matter (PM) Standards, United States, European Union, Japan, 1993–2010 (grams per kilowatt-hour)

Model Year	U.S. Trucks	U.S. Buses	European Union[a]	Japan
1993	0.3	0.13	0.4	n.a.
1994	0.13	0.094	0.4	n.a.
1995	0.13	0.094	0.4	n.a.
1996	0.13	0.067	0.15	0.25
1997	0.13	0.067	0.15	0.25
1998	0.13	0.067	0.15	0.25
1999	0.13	0.067	0.15	0.25
2000	0.13	0.067	0.1	0.25
2001	0.13	0.067	0.1	0.25
2002	0.13	0.067	0.1	0.25
2003	0.13	0.067	0.1	0.18
2004	0.13	0.067	0.1	0.18
2005	0.13	0.067	0.02	0.027
2006	0.13	0.067	0.02	0.027
2007	0.013	0.013	0.02	0.027
2008	0.013	0.013	0.02	0.027
2009	0.013	0.013	0.02	0.027
2010	0.013	0.013	0.02	0.027

[a] Euro IV from 2005 and Euro V from 2008.

NOTE: n.a. = not applicable.

SOURCE: Daisho Yasuhiro, Waseda University.

nia had already adopted its own motor vehicle pollution control program before the U.S. national program came into being, and the state has sufficiently large a market that an independent distribution system could provide unique vehicles for the rest of the country.

California adopted performance standards for vehicle exhaust emissions in 1968, the first place in the world to do so. Since then, the standards have been gradually tightened. These performance standards have resulted in a gradual reduction in emissions from cars as newer cleaner vehicles have replaced older, more polluting ones. In fact, California has taken the lead in stimulating the development and mandating the commercial introduction of advanced zero emissions technologies, including electric and fuel cells, many of which also can improve fuel efficiency.

Standards also have been established in the United States that require that specified emissions levels be maintained under special geographic conditions. For example, vehicles must be able to meet standards at both sea level and at an altitude of 1,609 meters (m).

European Union

As the pollution control program has matured in Europe, stringent vehicle emissions standards have been adopted for all 15 member states of the European Union, and Norway and Switzerland, non-EU member states, have decided to adopt identical standards. Furthermore, some of the central and Eastern European countries that have applied for EU accession, such as Poland and the Czech Republic, have already moved to adopt the EU vehicle and fuel standards.

The European Union also includes a unique provision in its directives by which member states are allowed to encourage the early introduction of vehicles that meet future emissions standards or fuel standards through the use of tax incentives. These incentives have been used successfully in both Germany and Denmark.

Japan

Japan was the first major industrialized country to eliminate the use of lead in gasoline; it introduced stringent car standards requiring catalytic converters in 1978. Since then, it has gradually tightened gasoline-fueled car standards and most recently began to move rapidly to reduce diesel-fueled vehicle standards. Before the end of 2004 sulfur levels in diesel fuel will be lowered to a maximum of 50 ppm, and it is expected that all new diesel vehicles sold in 2005 will be equipped with particulate filters.

Like some European countries, Japan is encouraging tighter vehicle standards through the use of tax incentives.

With the enforcement of new emissions standards in the major member countries of the Organisation for Economic Co-operation and Development (OECD), substantial reductions in emissions will be occurring for all on-road vehicle categories, both gasoline and diesel. In addition, fuel sulfur levels will be limited to a maximum of 50 ppm or less.

Recent Developments around the World

Most regions of the world have been significantly tightening their motor vehicle regulations. The major recent developments are described in this section.

United States

- In 1998 the U.S. Environmental Protection Agency (U.S. EPA), in conjunction with the California Air Resources Board (CARB), imposed the largest enforcement action in history on the heavy engine industry.

• CARB formally decided in August 1998 that diesel particulate matter is a toxic air contaminant, triggering an effort to further reduce PM emissions from urban vehicles, including retrofit where feasible.

• During 1999 CARB took emissions standards to the next level, not only tightening carbon monoxide (CO), hydrocarbon (HC), NO_x, and PM requirements but also establishing the principles of fuel neutrality (diesel vehicles must meet the same standards as gasoline-fueled vehicles) and usage neutrality (light trucks and sport-utility vehicles used primarily as passenger cars must meet the same standards as cars).

• In December 1999 the U.S. EPA adopted light-duty vehicle standards closely modeled after the California LEV 2 (low-emission vehicle, so-called Tier 2) standards and tighter sulfur requirements for gasoline.

• In December 2000 the U.S. EPA tightened its heavy-duty engine emissions requirements, with a special focus on tighter PM and NO_x standards and on low-sulfur diesel fuel.

European Union

• During 1998 the EU adopted directives for light-duty vehicle emissions and fuel quality that tightened emissions standards significantly (2000 and 2005), broadened the scope of coverage (e.g., cold temperature), added several important features previously missing (onboard diagnostics, in-use durability), and imposed low sulfur requirements for diesel fuel and gasoline.

• The EU and the auto industry reached agreement in 1998 on a voluntary commitment to reduce by 2008 the CO_2 emissions per kilometer driven by 25 percent.

• In January 2000 the EU adopted the next phases of heavy-duty standards—European Emission Standard III (Euro III), IV, and V—which will likely result in particulate and NO_x aftertreatment.

Asia and Eurasia

• In 1999 Japan tightened its gasoline-fueled automobile standards for the first time in 20 years and introduced the next phase of diesel-fueled vehicle requirements. Also in 1999 Japan's Ministry of International Trade and Industry (MITI) and Japanese industry reached agreement about lowering CO_2 emissions from vehicles.

• During 1999 China formally adopted the Euro I auto emissions standards and decided to phase out the use of leaded gasoline across the entire country by 2000.

• Taipei adopted step four of its motorcycle control program in late 1999, effectively banning two-stroke motorcycles by the end of 2003.

• The Supreme Court of India banned the sale of leaded gasoline in Delhi as of September 1999 and mandated that all new cars meet Euro I auto emissions standards. Similar requirements were then phased in across the entire country in 2000. Delhi adopted Euro II standards in April 2000.

SUMMARY OF WORLDWIDE GOVERNMENT FUEL ECONOMY STANDARDS

In the United States, Western Europe, and Japan—the three major markets for light-duty vehicles—the policies for improving the fuel efficiency of these vehicles have evolved in sharply different directions. In the United States from the mid-1970s to the mid-1980s, the major focus was on the Corporate Average Fuel Economy (CAFE) program, whereby annual fuel economy standards applied to each manufacturer on average across its entire fleet of light-duty vehicles, subdivided into passenger cars and light-duty trucks, with separate requirements for each of these categories. During the same period, a "gas guzzler" tax was imposed on those cars with the poorest fuel economy. More recently, the government has emphasized shared research and development (the Partnership for a New Generation of Vehicles, to be followed in 2002 by the FreedomCAR program) and tax incentives for certain high-efficiency vehicles (proposed but not yet enacted). In Europe, the European Automobile Manufacturers Association (ACEA) has proposed, and the European Union has accepted, a voluntary agreement pledging to reduce per-vehicle CO_2 emissions by 25 percent between 1995 and 2008. And in Japan the national government has established a series of weight-class fuel economy standards that require about a 23 percent improvement in the fuel economy of gasoline-fueled light-duty vehicles by 2010. Each of these programs will be reviewed in the rest of this section.

United States: The CAFE Program

The United States has had a mandatory fuel efficiency program since 1975. The Energy Policy and Conservation Act (1975), which came into effect in model year 1978, amended the Motor Vehicle Information and Cost Saving Act to require new passenger cars to achieve at least 27.5 miles per gallon (mpg) or 8.55 liters per 100 kilometers (km) by 1985, as measured by U.S. EPA test procedures. Separate, and more lenient, CAFE standards were first applied to light-duty trucks, including jeeps and minivans, in 1979. The current standard, set in 1996, is 20.7 mpg.

In recent years, as fuel prices have dropped and CAFE standards have remained unchanged, vehicle manufacturers have sold a growing pro-

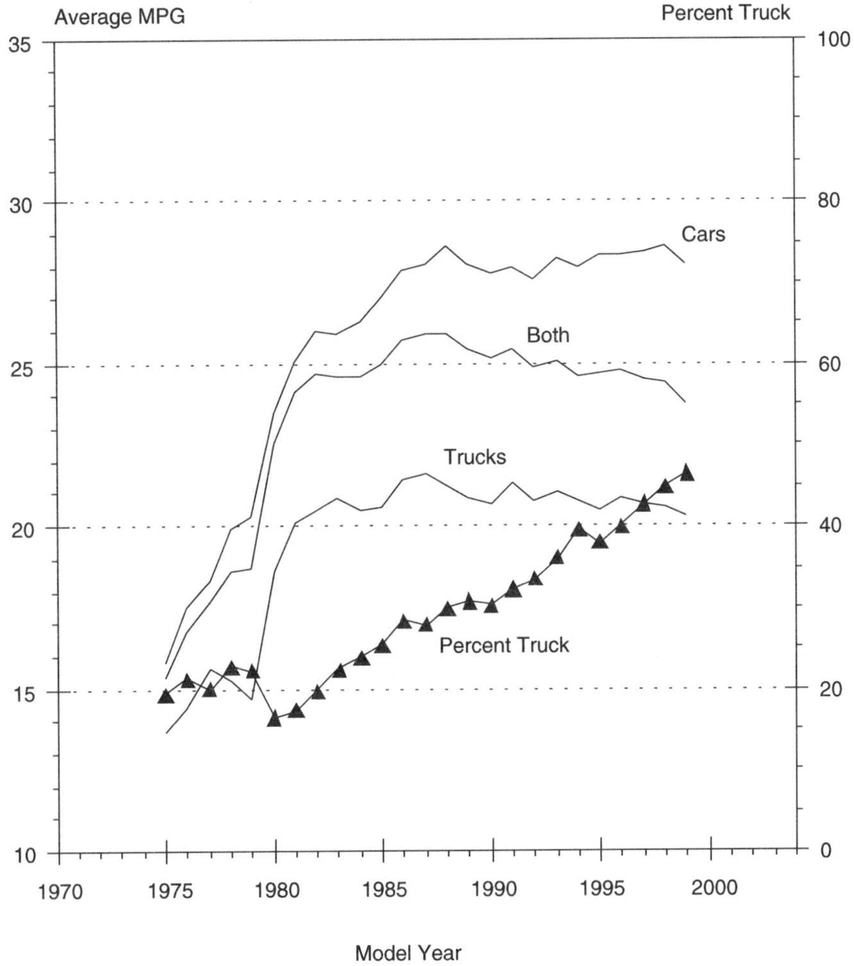

FIGURE 8-2 Miles per gallon (mpg) of trucks and cars by model year, United States. SOURCE: Hellman and Heavenrich (2001).

portion of light trucks. As a result, the overall fuel economy of new light-duty vehicles has been diminishing since 1980 (Figure 8-2).

Vehicle manufacturers are required to test a sample of all vehicles destined to be sold in the United States so that a fuel consumption rating can be assigned to each product line. The test involves both city and highway driving cycles. From these figures, a sales-weighted average fuel consumption figure is calculated for all the passenger cars produced by each manufacturer. Fuel efficiency (in miles per gallon) calculated this way must exceed the CAFE standard specified for the appropriate model year.

Failure to meet the CAFE requirements can result in substantial financial penalties. For each vehicle produced, a manufacturer whose fleet-average fuel consumption does not meet the CAFE standard is fined $5 per vehicle for every 0.1 miles per gallon by which the standard is not met. These fines may be offset by credits accrued in other model years, however. Since 1983 the federal government has collected $164 million in CAFE fines.

Another policy instrument used by the U.S. government is the gas guzzler tax, paid by people who buy cars with fuel economy below a certain threshold. The threshold rose from 5 mpg (12.1 liters per 100 km) in 1984 to 21.0 mpg (11.2 liters per 100 km) in 1985, and has been set at 22.5 mpg (10.1 liters per 100 km) since 1986.

Besides the CAFE requirements and gas guzzler tax, the federal fuel efficiency program provides consumers with information about the relative efficiency of new cars. The *Gas Mileage Guide* published by the U.S. EPA and the Department of Energy lists the city and highway fuel economy results of each vehicle model and is intended to provide information to new-car buyers. Also required on new cars are stickers indicating the vehicle's fuel economy as determined by the U.S. EPA, an estimate of the annual fuel cost based on 15,000 miles (24,000 km) of operation, and the range of fuel economy achieved by similar-size vehicles of other makes. U.S. EPA adjusts the measured fuel economy value downward before placing it on the sticker in an effort to give a somewhat more realistic estimate of the on-the-road fuel consumption that the owner can expect under average driving conditions. To allow a comparison of different vehicle models, U.S. EPA adjusts the mile per gallon estimates on all the new car stickers.

European Union

In the European Union, fuel economy is addressed by regulating CO_2 emissions. Carbon dioxide produced by passenger cars accounts for about half of CO_2 emissions from transport and about 12 percent of total CO_2 emissions in the European Union.[2] Under a "business as usual" scenario, CO_2 emissions from cars are expected to increase, from 1990 levels, by

[2] Derived from "A Community Strategy to Reduce CO_2 Emissions from Passenger Cars and Improve Fuel Economy," COM (95) 689, a communication from the Commission to the Council and the European Parliament. This strategy was adopted by the commission on December 20, 1995.

about 36 percent by the year 2010. The road transport sector has stood out in recent years as one of the few EU sectors experiencing growth in CO_2 emissions.

The European Union remains on track to achieve its long-standing commitment to stabilizing emissions of carbon dioxide—the main greenhouse gas responsible for manmade global climate change—at their 1990 level and then to reduce greenhouse gases by 2008. Total CO_2 emissions from the 15 EU member states were 0.5 percent lower in 2000 than 10 years earlier, according to the latest emissions inventory from the European Environment Agency (2002).

Less positive, however, is the fact that EU emissions of carbon dioxide and other greenhouse gases rose between 1999 and 2000, the most recent year for which EU-wide data are available. If transport were not included, there would have been a clear downward trend in carbon dioxide, nitrous oxide, and methane emissions across the EU. However, over the decade nitrous oxide from transport about doubled and carbon dioxide increased by about 20 percent.

In the face of these concerns, the European Automobile Manufacturers Association has entered into a voluntary agreement with the European Commission to reduce CO_2 emissions from new light-duty passenger vehicles, with firm fleet-wide targets of 140 grams (g) of CO_2 per kilometer (about 41 mpg for gasoline) by 2008, measured under the new European test cycle (Directive 93/116/EU). This represents about a 25 percent reduction from the 1995 average of 187 g/km (about 30 mpg) for this cycle. Because the European cycle is likely to produce lower fuel economy ratings than the U.S. combined city/highway cycle, the "U.S.-equivalent" miles per gallon ratings[3] of the year 2008 European fleet will likely be higher than 41 mpg if the targets are met.

Note that the goal of 140 g of carbon dioxide per kilometer is a collective target, not a target for each company. The participants in the agreement—BMW, Fiat, Ford of Europe, GM Europe, DaimlerChrysler, Porsche, PSA Peugeot Citroën, Renault, Rolls Royce, Volkswagen, and Volvo—have not publicly defined individual objectives, but before signing the agreement they discussed among themselves the likely trade-offs that would have to be made to achieve the goal. The agreement applies to light passenger vehicles classified as M1 in European Council Directive 93/116/EEC, which includes vehicles with no more than eight seats in

[3] Without discounting. The miles per gallon values that appear on new car stickers reflect a 10 percent discount off the city test results and a 22 percent discount off the highway test results, or 15 percent off the combined 55-city/45-highway rating.

addition to the driver. The agreement included a promise to introduce some models emitting 120 g/km (about 48 mpg) or less by 2000, and a nonbinding 2003 target range of 165–170 g/km (about 34–35 mpg). In addition, the commitment will be reviewed in 2003, with the aim of moving toward a fleet goal of 120 g/km by 2012. Finally, ACEA agrees to monitor compliance with the agreement jointly with the commission.

In exchange for its commitment to meeting the 2008 CO_2 emissions goal, the industry asked that some conditions be met:

- *Clean fuels availability.* Because the industry believes that direct-injection engines will play a key role in achieving the targets, the agreement asks for the "full market availability" of the clean fuels needed for this technology by 2005—gasoline with 30 ppm sulfur content and less than 30 percent aromatics, diesel fuel at 30 ppm sulfur and a Cetane number greater than or equal to 58.[4]
- *Protection against unfair competition.* Non-ACEA members must commit to similar goals, and the European Union will agree to try to persuade other car manufacturing countries to embrace equivalent efforts. The latter effort is designed to protect ACEA members from suffering in world market competition for their European efforts. Both the Japanese Automobile Manufacturers Association (JAMA) and the Korean Automobile Manufacturers Association (KAMA) have agreed to a revised version of the ACEA targets, with achievement of the 2008 target levels in 2009.
- *Regulatory cease-fire.* There will be no new regulatory measures to limit fuel consumption or CO_2 emissions.
- *Unhampered diffusion of technologies.* The companies assume that the commission will take no action that would hamper the diffusion of efficiency technologies, particularly direct-injection gasoline and diesel engines. Presumably, actions the commission might take could include tighter emissions standards on nitrogen oxides and particulates.

Japan

The Japanese government has established a set of fuel economy standards for gasoline- and diesel-powered light-duty passenger and freight vehicles, with fuel economy targets based on vehicle weight classes. The

[4] In mid-2000 the European Commission requested that all interested stakeholders comment on the benefits and penalties associated with the adoption of both gasoline and diesel fuel with a maximum of 30 ppm sulfur or near zero (< 10 ppm) sulfur compared with the currently mandated maximum of 50 ppm by 2005. As a result of this process, the commission has proposed the mandatory phase-in of fuels with a maximum of 10 ppm sulfur by 2005.

targets for gasoline-powered vehicles are to be met in 2010; 2005 is the target year for diesel-powered vehicles. The targets are to be met by each automaker for each weight class—that is, automakers cannot average across weight classes by balancing a less-than-target vehicle in one class with a better-than-target vehicle in another.

Compliance with these standards will produce by 2010 and 2005, respectively, a Japanese gasoline-fueled, light-duty passenger vehicle fleet capable of achieving 35.5 mpg and a light-duty diesel fleet able to achieve 27.3 mpg[5] as measured using the Japanese 10–15 driving cycle. The Japanese 10–15 driving cycle is substantially slower than the combined U.S. city/highway cycle, and the U.S. equivalent miles per gallon for this fleet would be significantly higher.

The regulations call for civil penalties if the targets are not met, but the penalties are very small. Realistically, enforcement will be accomplished through pressure from the government and the auto companies' desire to avoid public embarrassment, not through the financial penalties.

The fuel economy targets were selected by identifying "best-in-class" fuel economies in each weight class and demanding that the *average* new vehicle meet that level in the target year. The Japanese call this the "top runner" method of selecting fuel economy targets. Theoretically, this method is not "technology forcing" in that the technology has already been identified. Practically speaking, however, the standards may prove to be technology forcing because the "top runners" in each weight class must fully match their competitors in other areas of performance and amenities.

The fuel economy regulations have additional requirements over and above the actual fuel economy targets. These are:

- For new vehicles, fuel economy and major efficiency technologies on board must be recorded in catalogs and displayed at exhibits.
- Government is charged with providing education and other incentives for vehicle users and manufacturers, making sure that fuel economy regulation proceeds in harmony with other regulations (especially new emissions standards), reviewing manufacturers' efforts to improve fuel economy, and trying to harmonize this regulation with similar efforts in Europe and the United States.
- Manufacturers are expected to develop new efficiency technologies, design vehicles of outstanding efficiency, and help educate the pub-

[5] In the Japanese fleet light-duty diesel vehicles are larger, on average, than light-duty gasoline vehicles.

lic. It is assumed that the public will select efficient vehicles and use them in an efficient manner.

INSPECTION AND MAINTENANCE (I/M) PROGRAMS

Modern vehicles depend on properly functioning components to keep pollution levels low. Minor malfunctions in the air and fuel (A/F) or spark management systems can increase emissions significantly, while major malfunctions can cause emissions to skyrocket. A relatively small number of vehicles with serious malfunctions frequently cause the majority of the vehicle-related pollution problems. Unfortunately, it is rarely obvious which vehicles fall into this category, because the emissions themselves may not be noticeable and emissions control malfunctions do not necessarily affect vehicle drivability. Effective I/M programs, however, can identify these problem cars and assure their repair. The appendix to this chapter describes the elements of an inspection and maintenance program.

INDUSTRY–GOVERNMENT PARTNERSHIPS

The United States, Europe, and Japan have supported their domestic automotive industries in a variety of ways. This section focuses on government-industry research and development (R&D) partnerships in the United States and Europe.

Research covers a wide range of activities from basic research at universities and government laboratories, to applied research activities that may involve multidisciplinary and multiorganizational collaboration, to development work by industry to create new and improved technologies for use in automobile design and production. Although there is a spectrum of research activities ranging from very basic to very practical projects, the research process is not linear and often applied research will identify new opportunities for basic research and vice versa.

Basic research is aimed at acquiring knowledge of general importance and typically has a time horizon of a decade or more for success. Only a few basic research ideas may ever reach the marketplace. However, funding for basic research is frequently a small part of overall R&D budgets. Industrially supported efforts are often associated with institutions of higher learning. Funding sources include government, foundations, and collaborations with international groups. The kinds of research capabilities that will support a strong auto industry in China might include the following areas:

- clean fuels
- air quality monitoring and modeling techniques

- advanced propulsion systems (including fuel cells)
- catalysis and separations
- improved materials
- advanced electrical and electronic systems
- sensors and controls
- advanced manufacturing processes and systems.

Applied research includes a wide range of activities, including further refinement of promising innovations identified in basic research, feasibility and assessment studies, systems analysis and planning studies, and need-driven research investigations. Typically, applied research has a time horizon of 5–15 years and involves teams of researchers with different skills and backgrounds. The cost of such research is frequently greater than that of basic research by a factor of 10, but participation of government laboratories and industrial researchers expands the resource base. Examples of applied research projects are:

- innovative technology feasibility studies
- automotive system design and integration
- manufacturing systems
- infrastructure planning (roads, fuels, mobility services)
- traffic coordination and safety
- life cycle assessment of alternatives
- economic assessments.

Development and demonstration research activities are usually funded and conducted by industry for those technologies that appear to offer significant potential for near-term (3–5 years) commercial advantage. Significant segments of these activities are likely to be proprietary. Because large investments are needed, such projects are selected very carefully. Often they will involve collaborative work between a key component supplier and system designers. These activities also may present an opportunity among joint venture partners for personnel exchanges that will broaden the capabilities of both partners. Because of the investment requirements, some companies prefer to wait for others to make the breakthrough and then either purchase rights to the technology or adapt it to their needs. In areas in which large investment is needed to develop a new product that may not be a near-term market success (e.g., the U.S. PNGV program discussed later in this chapter), government-industry partnerships may fund the research, spreading the investment required over a number of sponsors. Here the final product may not be a commercial success, but the research will produce know-how and component technologies that may be well worth the investment.

The cost of building a strong auto research capability in China will be large. Although China may wish to let others take the lead in the longer-term research activities, it will have to maintain the capability to stay abreast of what is being done worldwide and limit its investment to areas in which a particular concept seems to offer a special advantage to the Chinese industry or in which the educational benefits are worth the investment. However, when commercial success is the end goal, it is usually best to let the industry make the choices about technologies. Government and academic researchers may offer ideas and guidance, but only the industry has the important knowledge about how individual technologies integrate into a successful car.

United States

Early Efforts

One of the earliest efforts to undertake industrial cooperative research was organized through the Inter-Industry Emission Control Program in 1972. This program was a joint effort of members of the U.S. petroleum industry, members of the Japanese automotive industry, and the Ford Motor Company. Other members of the U.S. automotive industry were prevented from participating by antitrust laws. The program, which was funded by the various participants, continued for over a decade and produced some significant technical developments. It was terminated when the emphasis of the industry turned more to fuel economy than to meeting selected emissions standards.

In the late 1970s Secretary of Transportation Brock Adams initiated a government effort to "reinvent the automobile" as the technological answer to the influx of high-efficiency small cars from Japan. The result was a government-industry program designed to emphasize basic research that would enhance the efficiency of vehicles. Annual joint funding of the Cooperative Automotive Research Program was pegged at about $100 million. All parties had signed the agreement and were preparing to launch the program when it was cancelled by the newly elected Reagan administration in 1981.

Background of the Partnership for a New Generation of Vehicles (PNGV)

In the late 1980s Congress began to restrict the ability of the National Highway Traffic Safety Administration to adopt tighter CAFE requirements, and fuel economy standards were effectively frozen. Seeking other routes toward fuel economy, the major U.S. automakers and the U.S. government, through its national laboratories, began sharing technology information and manufacturing know-how in 1993 in an effort to develop low fuel consumption technologies.

On September 29, 1993, President Bill Clinton and the chief executive officers of the major domestic automakers (Chrysler, Ford, and General Motors) announced the formation of the Partnership for a New Generation of Vehicles. The long-term goal of PNGV was to develop vehicles that would deliver up to three times the current fuel efficiency (defined as 80 mpg or energy equivalent) and would cost no more to own and operate than the current comparable vehicles. At the same time, this new generation of vehicles was to maintain the size, utility, and performance standards of contemporary vehicles (i.e., the 1994 Chrysler Concorde, Ford Taurus, and Chevrolet Lumina) and meet all mandated safety and emissions requirements.

The U.S. automobile manufacturing industry is an integral part of the U.S. economy, accounting for one out of seven U.S. jobs and 4.5 percent of the gross domestic product. The development of a new generation of vehicles was to improve U.S. competitiveness by establishing the capability for technical leadership in the production of competitively priced, highly fuel efficient, low-emission automobiles. Improvements in advanced manufacturing techniques that shorten product development times and lower costs, as well as improve product quality and durability, are essential to transfer new technologies to the marketplace affordably. The PNGV represented a departure from the historical, primarily regulatory relationship between government and the U.S. automobile industry. Because the current U.S. price of gasoline did not encourage consumer demand for high-efficiency automobiles, government support of long-term research and development for fuel efficiency technologies was considered necessary to spur activity and accelerate progress in the absence of market pull.

The achievements of the PNGV were expected to produce significant energy, environmental, and economic benefits for the nation. In view of the country's growing population and Americans' fondness for travel, a significant improvement in vehicle fuel efficiency would be a major step toward lessening reliance on foreign oil supplies and reducing the associated balance-of-trade deficits, which were greater than $40 billion in 1993.

PNGV was structured to achieve three mutually supportive, interactive goals:

1. Significantly improve national competitiveness in manufacturing for future generations of vehicles.

2. Improve the productivity of the U.S. manufacturing base by significantly upgrading U.S. manufacturing technology, including adopting agile and flexible manufacturing processes and reducing cost and lead

times, while also reducing negative environmental impacts and improving product quality.

3. Implement commercially viable innovations from ongoing research in conventional vehicles. Pursue technology advances that can lead to improvements in fuel efficiency and reductions in the emissions of standard vehicle designs, while pursuing advances to maintain safety performance.

Research focused on technologies that reduce the demand for energy from the engine and drive train. Indeed, the auto industry pledged to apply those commercially viable technologies resulting from this research that could be expected to significantly increase vehicle fuel efficiency and improve emissions.

As noted, the objective was to develop vehicles that could achieve up to three times the fuel efficiency of comparable 1994 family sedans—the Concorde, Taurus, and Lumina automobiles—with equivalent cost of ownership, and yield a revolutionary class of fuel-efficient, environmentally friendly, commercially viable vehicles that would meet or exceed safety and emission requirements. The PNGV's target was to develop a concept vehicle by 2000 and a production prototype by 2004. This 10-year time frame for the PNGV represented a rapid development effort to produce a revolutionary change in automotive transportation.

In 1994 the auto industry identified the areas in which significant innovations were needed to meet PNGV goals: reduced vehicle weight, more efficient power trains, and reduced parasitic losses. A critical element in meeting the technical challenges in these areas was believed to be the development of manufacturing processes capable of quickly delivering high quality and volume at low cost.

Participants in the PNGV Technical Program

The Partnership for a New Generation of Vehicles was formed by the federal government and the U.S. Council for Automotive Research (USCAR), which represented the major American auto companies—Chrysler (which became DaimlerChrysler in 1998), Ford, and General Motors. The original government members of the partnership were the Departments of Commerce (designated as lead agency), Defense, Energy, Interior, and Transportation; Environmental Protection Agency; National Aeronautics and Space Administration; and National Science Foundation. Other participants in the PNGV R&D activities included industrial suppliers, universities, commercial R&D institutions, and entrepreneurs.

The PNGV organization was overseen by a steering committee of senior representatives from the three automakers and the Department of Commerce, with a rotating director. One level below the steering committee was a technical committee composed of representatives of the companies and seven government agencies. Under the technical committee were some 10 technical working groups for the major technology subsystems, staffed by engineers and scientists from industry and national labs. Most of the groups were chaired by an industry representative.

The PNGV was able to pursue an ambitious program schedule by leveraging ongoing government and industry R&D programs. Before announcement of the PNGV, the federal government and the automotive manufacturers and their suppliers had already launched cooperative research programs, and additional government-sponsored research was already being conducted in government laboratories and universities. Technology development was particularly successful in the areas of innovative power trains (hybrid vehicles, gas turbines, and fuel cells), lightweight materials (structural aluminum and magnesium and various composites), and energy storage devices (such as ultracapacitors, flywheels, and batteries). Many of these ongoing R&D programs provided the "running start" considered vital to achievement of the PNGV goals within the allotted time.

As a result of the long lead time in federal budgeting automaker and federal laboratory managers shifted a variety of existing vehicle R&D projects to the PNGV program, including about $250 million in multiyear hybrid vehicle research already in place within Ford and General Motors. The U.S. General Accounting Office estimates that federal support for the partnership averaged about $250 million a year from 1995 through 1999, but this sum overstates support for the partnership itself because about 45 percent supported activities only indirectly relevant to the partnership goals or was not coordinated through the partnership (U.S. GAO, 2000).

In addition to government-assisted research and development, automotive manufacturers maintain both proprietary and nonproprietary programs in advanced technology research in order to assure their competitive positions. Proprietary research contributions increased as the PNGV program moved through the development of concept cars and production prototype vehicles. Indeed, it was reported that industry was matching government funds with about $250 million a year, but in fact "a major portion" of the spending was in proprietary product programs (NRC, 2001:10).

In the early years, of the some $293 million a year the government was spending on PNGV, about a third went directly to the federal laboratories, about a third directly to automotive suppliers, and about a third to the three automakers. Of the third that went to the three automakers, about three-quarters was later subcontracted to suppliers (Chapman, 1996). The three automakers may have received a relatively modest amount of money, but they played a large role in determining how the money was spent and by whom.

In 1994 the federal government asked the National Research Council to establish an independent standing committee to prepare an annual review of the PNGV program. Committee members, many of whom had automotive backgrounds, were experts on different aspects of the program. Seven reports were issued.

Evaluation of the PNGV Program

In 1997, as planned, the large set of candidate technologies that had been examined during the first years of the partnership was reduced to a few for further development. Each of the three companies selected diesel-electric hybrids as their preferred technology. In early 2000, again in line with program milestones, each of the three companies unveiled concept cars. Ford's Prodigy, GM's Precept, and DaimlerChrysler's ESX3 all used lightweight materials and combined small advanced diesel engines with electric drive trains, with projected fuel economy of 60–80 mpg (NRC, 2001). As indicated in the seventh (and last) annual review of the PNGV program by the National Research Council, the automotive companies appeared to be meeting the program schedule for achieving the fuel economy goals, but they would not meet the cost goals (NRC, 2001). The efforts to meet the emissions goals are discussed later in this chapter.

The National Research Council's seventh review of the PNGV Program made the following observations about the achievement of program goals (NRC, 2001):

> At the end of 1997 PNGV made a technology selection based on assessments of system configurations for alternative vehicles. Several technology options—such as gas turbines, Stirling engines, ultracapacitors, and flywheels for energy storage—were eliminated as leading candidates. The 10-year span of the program dictated some of these choices. In its fourth review the committee agreed with PNGV's technology selections (e.g., four stroke, internal combustion engines, fuel cells, batteries, power electronics, and structural materials). The four-stroke compression-ignition direct-injection (CIDI) engine was selected as the most likely power plant to enable the fuel economy goal to be met within the program time frame; the fuel cell power plant was retained in the program as a highly promising longer-range technology.
>
> * * *
>
> The second major milestone, the development of concept vehicles, was met in early 2000. Using PNGV-developed technologies and their own in-house proprietary technologies, the . . . . companies each developed separate concept vehicles with fuel economies between 70 and 80 mpg.
>
> * * *
>
> The power train with the highest probability of meeting the vehicle fuel-

> economy target of 80 mpg by 2004 is the hybrid-electric power train powered by a CIDI engine. In 1999 approximately midway through the program, the Environmental Protection Agency promulgated Tier 2 emission standards for particulate mater and oxides of nitrogen (NO_x) substantially more stringent than those at the start of the program. . . . This action brought into question the possibility of meeting these emission requirements with a CIDI engine in a production prototype by 2004. Consequently, a major portion of the program resources was reallocated to address this new development risk. Alternative power plants (e.g., homogeneous spark-ignition engines or gasoline-fueled direct-injection engines) with a higher probability of meeting the Tier 2 standards in the PNGV 2004 time frame would result in vehicles with reduced fuel economy compared with the CIDI engine.

Perhaps equally important, the program gave rise to a "boomerang effect"—that is, the existence of this program encouraged competitors to go forward more aggressively (Sperling, 2001). Apprehensive European and Japanese automakers quickly accelerated their efforts through programs such as the European Car of Tomorrow Task Force (1995) and the Japan Clean Air Program (1996), and through individual company efforts such as Toyota and then Honda's commercialization of hybrid electric cars and Daimler-Benz's enhanced fuel cell program. Many executives in European and Japanese companies readily concede that PNGV was clearly seen as a threat, and that it therefore served as the catalyst for increased investment in advanced propulsion technology in their companies. The competition intensified as U.S. automakers responded to the aggressive commercialization efforts by Toyota, Honda, and the Daimler side of DaimlerChrysler. As for gauging the success of the program, one might ask: Why did the PNGV effort not lead to the commercial advances envisioned in 1993, even in some cases when those advances were being actively pursued by automakers and other technology companies elsewhere the world? As noted earlier, the National Research Council evaluation suggests that the shortcoming stemmed from the initial schedule and design of the program (NRC, 2001). Indeed, it appears that PNGV formalized a very ambitious schedule with specified deliverables that led, ironically, to a conservative approach. Fearing that the time horizon that was too short to allow much development of emerging technologies, industry and government managers focused on relatively mature technologies for which fuels were available—that is, diesel-electric hybrid cars. Even then, automakers were falling far short of meeting the goal of comparable cost.

Another major issue for the PNGV program, and government-industry partnerships in general, was control of knowledge and rights to technology. The automakers, adhering to common practice in competitive industries, essentially created "firewalls" of varying permeability around their PNGV work. Companies engaging in collaborative work with competitors in their own or related industries routinely create these walls to protect themselves against antitrust lawsuits and, more important, to ensure confidentiality. The concern is that the more government funding and competitors are involved, the more likely it is that companies will lose control of knowledge and technology.[6] These firewalls work effectively with small innovations that affect a small part of the business, because the protected knowledge may not be central to the business interests of the company. But this situation was different. First, virtually all of the targeted technologies were close enough to commercialization that a company would want proprietary rights to any advances. Second, fuel cell and hybrid propulsion systems, if successfully developed, had the potential to be core technologies for these huge companies.

In any case, the PNGV experience provided the following insights and lessons. First, if properly structured, a joint program between government and industry can successfully develop new technologies of interest to the commercial sector. Second, unforeseen indirect effects (the "boomerang effect" in this case) may prove to be very important. Third, program objectives must be reexamined in light of changing conditions and objectives changed accordingly. Fourth, targeted technologies should be far from commercialization (or have large social benefits). Fifth, progress is greatest with partners wholly committed to the technology development and commercialization goals of the partnership. Finally, it is important that the government limit its participation to the noncompetitive phase of research and development and leave the final development of a marketable product to industry.

Overall, the partnership generated many successes. An important benefit has been the greater communication between industry and government and therefore less adversarial tension. The high-profile collaboration between government and the automotive industry also spurred the development of new technologies, many of which are being used to improve the efficiency of vehicle subsystems and components. The program also focused government's advanced technology R&D efforts and highlighted, for the public and the automotive industry, the potential for major technology enhancements.

---

[6] The financial and legal claims by government on publicly funded innovations vary greatly. The European Union, for example, rarely asserts a claim to technologies developed by automakers with public funds. The U.S. government, in contrast, has become quite aggressive at asserting a claim.

The Next Phase of Cooperation

In January 2002 the U.S. government announced a significant redirection of the PNGV program and gave it a new name: the FreedomCAR (Cooperative Automotive Research). Rather than maintaining the heavy focus on demonstration vehicles, the program will concentrate more on technology and on developing the fuel cell for passenger vehicles and a hydrogen fuel infrastructure for those vehicles. The generation of hydrogen fuel, its distribution and storage, and its storage or generation on board the vehicle, will be an important part of the program. A particular thrust of the program will be an effort to ensure more coordination among its various participants—industry suppliers, universities, and government laboratories. USCAR will continue to be the sole industry partner, and the Department of Energy will serve as the lead government agency. However, an effort will be made to expand the membership to include energy suppliers, along with automakers, their suppliers, and research groups from around the country. A procedure for external review of the program will continue, but on a biannual basis.

The European Experience

In Europe, automakers and governments have engaged in several high-profile international R&D partnerships since the mid-1980s.[7] The most recent incarnation, the European Council for Automotive R&D (EUCAR), was launched in 1994, partly in response to the PNGV program. EUCAR comprises 10 automotive companies located in five countries and has its headquarters in a sixth, Belgium. From 1994 to 1999 EUCAR undertook 88 projects, of which 14 were self-funded (by EUCAR members) and 74 were cofunded with the European Union. The total budget for the 88 projects was EUR302 million, about half of which was provided by the European Union (half of that going to the automakers and the other half to suppliers, universities, and independent centers). EUCAR has created an array of technical and policy committees not unlike those associated with the PNGV program, which are aided by a skeletal administrative staff.

Interviews with a variety of senior officials from the government and automotive partners indicate that two major benefits have arisen from the EUCAR partnership: (1) automakers have gained access to European research institutes (which are similar to the U.S. national energy laborato-

[7] This section is based on Sperling (2001).

ries); and (2) communication has increased across the industry and between industry and the European Union. EU funding itself was rarely cited as an important benefit. The EU provides even less public R&D funding to automakers than that provided through the PNGV program.

EUCAR, then, is principally an organization designed to share information. With the challenge of managing the politics and interests of a wide variety of countries and a broader array of companies, the "cultural" commitments of some countries to their major car companies, and the various relationships among governments, universities, and companies, it is difficult to imagine EUCAR expanding into an integrated R&D partnership. EUCAR also has played a pivotal role in maintaining communication between the European Commission (the executive arm of the European Union) and automakers about follow-up to their voluntary agreement to reduce CO_2 emissions (per vehicle-kilometer) by 25 percent between 1998 and 2008.

General Lessons of PNGV and EUCAR

The PNGV and EUCAR programs can provide valuable insights and lessons for China. Partnerships can play an important role in identifying technologies that are in the public interest and also commercially viable. Moreover, they help inform the public debate about new vehicle technologies, highlight opportunities, and provide a mechanism for directing government resources. Independent research centers play an important role in accelerating development, and such research centers will be most successful when closely partnered with industry members. Government funding of universities is critical, and a principal aim should be training of engineers and scientists.

As any such partnership proceeds, the goals and programs need to be flexible and reviewed on a regular basis. The government must recognize that the companies will need to control any proprietary knowledge they develop. All companies will expect to be treated equally.

In a free market economy, decisions about commercialization must remain with the industry. Although the commercialization of advanced technology is most likely to occur in response to specific performance standards and goals or competitive market forces, R&D "partnerships" can provide important information during the pre-competitive phase of the development process. An important lesson from the PNGV experience for China is related to developing human capabilities. Even in the United States, with its massive university and national lab systems, the automotive companies concluded that "the lack of talented people is a greater handicap than the lack of adequate funding and [we] need ideas (breakthroughs) more than dollars" (NRC, 2000:9)

In the end, China will need to adjust any strategy it follows for generating and sharing knowledge and working with industry to fit its special circumstances. Such a strategy will differ from the U.S. PNGV experience because China does not have large, existing automotive-related research capabilities in its industry, universities, or government research centers. One overarching lesson learned, however, is that the partnership process, in whatever form it takes, is difficult and requires a strong commitment on both sides. Also, because reducing energy consumption and emissions is a large-scale systems problem, it is important that all of the key players be involved in any partnership process. One weakness of the PNGV program was that the energy suppliers were not partners in the effort. The overriding lesson, though, is that in this globalizing and networking world, communicating and partnering are more essential than ever. Because foreign original equipment manufacturers are highly involved in the Chinese automotive industry, the government must soon decide how much these foreign members will be allowed to participate in any government-industry program.

CONCLUSION

Governments have an important role to play in fostering improved research and development and an even more important role in determining the attributes of individual vehicles. The government's role in research and development can vary from providing resources for basic research to stimulating the development of human resources. As for vehicle attributes, government can stimulate advanced technologies by setting performance standards or introducing strong incentives for rapid advances. It is important that those trying to leapfrog to more advanced technology pay careful attention to other important conditions that could limit their success such as assuring the availability of fuels of the appropriate quality.

APPENDIX: INSPECTION AND MAINTENANCEPROGRAMS

Effective inspection and maintenance programs can identify the cars with emission control malfunctions and assure their repair.[8] Test procedures must keep pace, however, with the advances in vehicle technology.

---

[8] Some of the material in this appendix has been derived from LAT, Aristotle University of Thessaloniki, Greece (1998). The project described in this report was funded by the European Commission, Directorate Generals for Environment (DG XI), Transport (DG VII) and Energy (DG XVII).

For the most advanced vehicles, the emissions, when properly maintained, will be so low that a more sophisticated test will be required. Vehicles equipped with electronic controls of air-fuel and spark management systems and equipped with catalytic converters to reduce CO, HC, and NO_x emissions are best tested using a transient test on a dynamometer that includes accelerations and decelerations typical of actual driving.

In general, I/M programs are most effective when they take the form of centralized I/M systems in which the testing of vehicles is completely separated from those carrying out repairs. These programs also cost much less overall because they tend to be high throughput.

The rest of this appendix summarizes the various test procedures that can be used in I/M programs and some of the more recent experiences with vehicle inspection and maintenance efforts.

No-Load Short Tests

The term *no-load* denotes all tests during which no external load is imposed and the car operates with the transmission in the neutral position.

Idle/Fast Idle Test

This test measures CO, HC, and CO_2 concentrations in the raw exhaust gas at idle speed and possibly a higher engine speed, 2,000–3000 revolutions per minute (rpm). The test could last from less than one minute for a one-speed idle test without preconditioning to about 10 minutes for a two-speed test that includes a "second chance" test with preconditioning (Tierney et al., 1991; Laurikko, 1994). A garage-type nondispersive infrared (NDIR) analyzer capable of measuring CO, HC, and CO_2 concentrations is sufficient for detecting the level of the pollutants.

Today, idle/fast idle tests are still widely used in I/M programs because they are the fastest, cheapest, and easiest to perform with the least possible testing equipment. For carbureted cars, they can effectively identify malfunctioning mixture preparation systems by checking the performance of the carburetor's idle mixture orifice in the idle test and the main fuel metering orifice in the fast idle test. However, modern cars equipped with electronic fuel injection and ignition systems and three-way catalysts may have a defect—such as defective sensors and degraded catalyst efficiency (Pidgeon and Dobie, 1991)—that cannot be detected through their pollutant emissions at idle; even worse, the great bulk of emissions may be generated during transient engine operation. An additional very significant drawback is the negligible amount of NO_x emissions at idle.

Idle/Fast Idle Test with Lambda Test

For catalyst-equipped cars, a lambda test may be coupled with an idle/fast idle test in order to check the performance of the mixture preparation system. Three steps are usually performed:

1. The fuel/air ratio is indirectly determined by measuring the CO_2, CO, O_2, and HC concentrations in the raw exhaust at fast idle (2,000–3,000 rpm).
2. The fuel/air ratio is artificially modified by adding oxygen, propane, or recirculated exhaust gas to the intake air, and the response of the lambda control system is checked. Long response times imply that the oxygen sensor is degraded, and no response means that the lambda control system is not operating.
3. One or more of the characteristics of the electronic lambda control circuit are measured and compared with the auto manufacturers' specifications.

Since December 1993, Germany has used a test that involves both test types 1 and 2. Evaluations have shown that the test performs fairly well with excess emitters. A combined idle/fast idle–lambda test (involving lambda test types 1 and 2) also is being used in Austria, where it has demonstrated satisfactory effectiveness (Pucher and Lenz, 1990).

Steady-State Loaded Tests

Because NO_x emissions at no-load conditions are negligible, a loaded test is required to measure NO_x emission levels, which are a critical source of urban air pollution. The simplest loaded tests involve a dynamometer with steady-state power absorption. A simulation of the car's inertia weight is not required, because there is no transient phase in the emission test: the car is driven at constant speed and load, and pollutant concentrations (CO, HC, NO_x, and CO_2) are measured during the load phase.

In response to the introduction of three-way catalyst-equipped cars, the acceleration simulation mode (ASM) test was developed. For this test the car is driven on a chassis dynamometer at a constant speed and steady-state power absorption that is equal to the actual road load of the car during acceleration. Thus one can achieve a realistic simulation of the car's load at a specific driving mode without the need of flywheels for inertia simulation. However, at high speed/high acceleration combinations the required power absorption is too high to be achieved without overheating the engine (Austin and Sherwood, 1989). Pollutant concentrations (CO, HC, NO_x) are in principle measured in the raw exhaust. Each steady-state test mode requires about 10 minutes for preparation, preconditioning, actual testing, and documentation.

Austin and Sherwood (1989) compared several ASM speed/load combinations with idle tests and already developed steady-state loaded tests as well as with a transient loaded test. The best results were obtained from the ASM 5015 test, which has a constant speed of 15 mph (24 kilometers per hours and a steady-state load equal to 50 percent of the load required to accelerate at 1.47 meters per second squared (m/s^2)—the maximum acceleration rate on the Federal Test Procedure (FTP)—at a speed of 15 mph.

In the late 1980s TÜV, a German company that undertakes a great deal of government-type work, including certification of new vehicles, I/M testing, and government research projects such as emissions factor tests, investigated a similar loaded test. The car is driven at 50 kph and at 7-kilowatt dynamometer power absorption in third gear (position "D" for cars with automatic transmission) and then idles; pollutant concentrations (CO, HC, NO_x) in the raw exhaust are measured at the end of both the loaded and the idling phases (Voss et al., 1987). Vehicle preparation, preconditioning, testing, and documentation take about 10 minutes. The study concluded that the test is much more appropriate than a simple idle/fast idle test for inspecting catalyst cars.

Transient Loaded Tests

In transient tests, cars are driven on the dynamometer according to a specific driving schedule; the main differences between these tests and those used for type approval or new vehicle certification are the duration of the driving cycle and the hot start. Because exhaust gas emissions are expressed in mass units, a constant volume sampler (CVS) system and laboratory-quality analyzers are required to detect low pollutant concentrations in the diluted exhaust sample. A multiple-curve dynamometer with flywheels also is required to simulate the instantaneous road load and the power needed to accelerate the inertia masses of each car.

The CDH 226 test developed by the Colorado Department of Health (CDH) sought to achieve a high correlation with the U.S. Federal Test Procedure, especially for three-way catalyst cars. Numerous studies have demonstrated correlation coefficients (R^2) of 0.79–0.96 for all three pollutants (Ragazzi et al., 1985; Austin and Sherwood, 1989; Klausmeier, 1994). Excess emission identification rates were about 90 percent for all three pollutants at 5 percent errors of commission (Ragazzi et al., 1985).

The U.S. Environmental Protection Agency, however, decided to develop a more transient alternative to the CDH 226 in order to better simulate the FTP. The result was the IM240 (Pidgeon and Dobie, 1991). The

IM240 procedure requires a constant volume sampler and laboratory-grade analyzers for carbon monoxide, hydrocarbons, nitrogen oxides, and carbon dioxide. Emissions in the diluted exhaust gas are normally derived on a mass basis with a CVS, and the test takes about 10 minutes to perform. The IM240 showed correlation coefficients of 0.89–0.97 for all three pollutants with the FTP hot start portion; another test sample showed coefficients of 0.54–0.82 with the full FTP, including cold starts (Klausmeier, 1994). This procedure evolved into the VMass test procedure, which has demonstrated very close correlation with the IM240 test but at much lower cost.

REFERENCES

Austin T. C., and L. Sherwood, 1989. Development of Improved Loaded-Mode Test Procedures for Inspection and Maintenance Programs. SAE Technical Paper No. 891120. Society of Automotive Engineers.

Chapman, R. 1996. Testimony, Hearing on Partnership for a New Generation of Vehicles (PNGV): Assessment of Program Goals, Activities and Priorities. Subcommittee on Energy and Environment of the Committee on Science, U.S. House of Representatives. 104th Cong., 2d sess. Washington, D.C.: Government Printing Office.

European Environment Agency. 2002. Annual European community greenhouse gas inventory 1990–2000 and inventory report 2002. Technical report no. 75. Online. Available at reports.eea.eu.int/technical_report_2002_75/en. Accessed October 22, 2002.

Hellman, K. H., and R. M. Heavenrich. 2001. Light-Duty Automotive Technology and Fuel Economy Trends. EPA 420-R-01-008. Advanced Technology Division, Office of Transportation and Air Quality, U.S. Environmental Protection Agency. September.

Kenworthy, J. R., and F. B. Laube. 1999. An International Sourcebook of Automobile Dependence in Cities, 1960–1990. Boulder: University Press of Colorado.

Klausmeier R. 1994. Analysis of I/M test alternatives. Paper presented at the International Conference on Ozone Control Strategies for the Next Decade (Century), San Francisco.

LAT, Aristotle University of Thessaloniki, Greece; INRETS, France; TNO, The Netherlands; TÜV, Rheinland, Germany; and TRL, United Kingdom—in collaboration with MTC, Sweden; IVL, Sweden; VKM-Thd, Graz University of Technology, Austria. 1998. The Inspection of In-Use Cars in Order to Attain Minimum Emissions of Pollutants and Optimum Energy Efficiency. May.

Laurikko J. 1994. In-Use Vehicle Emissions Control in Finland: Introduction and Practical Experience. SAE Technical Paper No. 940930. Society of Automotive Engineers.

National Research Council (NRC). 2000. Review of the Research Program of the Partnership for a New Generation of Vehicles: Sixth Report. Washington, D.C.: National Academy Press.

———. 2001. Review of the Research Program of the Partnership for a New Generation of Vehicles: Seventh Report. Washington, D.C.: National Academy Press.

Pidgeon W. M., and N. Dobie. 1991. The IM240 Transient I/M Dynamometer Driving Schedule and the Composite I/M Test Procedure. U.S. EPA Technical Report AA-TSS-91-1. U.S. Environmental Protection Agency, Washington, D.C.

Pucher E., and H. P. Lenz. 1990. First results in testing in-use catalyst cars using air-fuel ratio measurement from the exhaust gas. ATA 905114. Paper presented at the Twenty-third FISITA Congress.

Ragazzi, R. A., J. T. Stokes, and G. L. Gallagher. 1985. An Evaluation of a Colorado Short Vehicle Emission Test (CDH-226) in Predicting Federal Test Procedure (FTP) Failures. SAE Technical Paper No. 852111. Society of Automotive Engineers.

Sperling, D. 2001. Public-private technology R&D partnerships: Lessons from U.S. Partnership for a New Generation of Vehicles. Transport Policy 8(4):254.

Tierney E. J., E. W. Herzog, and L. M. Snapp. 1991. Recommended I/M Short Test Procedures for the 1990s: Six Alternatives. U.S. EPA Technical Report AA-TSS-I/M-90. U.S. Environmental Protection Agency, Washington, D.C.

U.S. General Accounting Office (GAO). 2000. Results of the U.S.-Industry Partnership to Develop a New Generation of Vehicles. GAO/RCED-00-81. Available at www.gao.gov.

Voss H.-J., D. Hassel, H.-P. Neppel, A. Richter, N. Heckötter, and A. Friedrich. 1987. Periodic Inspection of Exhaust Emissions from Low Pollutive Otto Engine Vehicles and Diesel Engine Vehicles. SAE Technical Paper No. 871084. Society of Automotive Engineers.

9

Findings and Recommendations

This chapter summarizes the findings of this report, and, based on those findings, makes specific recommendations for actions the committee believes are needed to build a competitive automotive industry in China while minimizing its adverse impacts on China's environment, society, and economy.

PATTERNS OF MOTORIZATION

Chapter 2 describes how China's pattern of motorization is similar thus far to that of many other countries, and how the size of its motor vehicle fleet is strongly associated with its growing income level. Similarly, the fact that paved road length increases as national income increases parallels the experience of other countries. If China's gross domestic product (GDP) continues to grow, these patterns indicate that the number of vehicles will continue to increase and that the automobile share of the fleet will increase greatly. However, the attributes of the vehicles will depend greatly on national and regional government policies. Among the most important are those for land use, traffic management, and transportation infrastructure, as well as for fuel prices and vehicle performance standards for emissions, safety, and energy efficiency. Countries with high fuel taxes tend to have a preponderance of smaller cars, and those with low fuel prices have larger cars. Rigorous emissions standards tend to promote higher technology and more expensive vehicles.

Recommendation

- *To pursue the goals of energy conservation and environmental quality, the Chinese government should employ vehicle performance standards on emissions, fuel economy, durability, and safety. Such a move is likely to restrict the field to those manufacturers that have the level of technology to meet the performance standards and are capable of producing vehicles of the required quality. The overall effect will likely be consolidation of the industry, as suggested in the five-year plan for the development of the automotive industry.*

The growing market for automobiles in China will require expansion of related services such as fuel distribution and sales, credit systems, automobile maintenance facilities, vehicle inspection, parking facilities, and liability insurance. Because most of these services are in the public sector in China, the government will have to ensure that services and facilities keep pace with market growth and do not unintentionally limit the acceptance of personal automobiles among the Chinese people.

THE FUTURE OF THE CHINESE AUTOMOTIVE INDUSTRY

After many years of effort, China has developed a strong domestic motor vehicle industry, but one that is heavily dependent on overseas manufacturers for technology. At present, China is the ninth largest car and truck producer in the world and the largest motorcycle manufacturer. Although its largest producers are partnered in joint ventures with several major world manufacturers, most of the technologies currently used in Chinese-manufactured vehicles are incorporated into the vehicle designs by the joint venture partners.

In its five-year plan for the automotive industry the Chinese government has laid out an ambitious program of industry consolidation in order to produce cars that will be competitive in international markets and will not depend on joint venture partners for key technologies. The plan sets clear targets for emissions control, safety, and fuel efficiency. And it endorses research efforts on advanced technologies such as electric, hybrid, and fuel cell power sources. But China faces many challenges in its effort to produce an indigenous vehicle that is clean, efficient, affordable, and internationally competitive. The major barriers are:

- The investment by domestic manufacturers in research and development (R&D) is far below that of foreign competitors.
- Even though China's car market has grown rapidly, it is still many

years away from being large enough to provide domestic car manufacturers with revenue sufficient to support a major R&D effort.

- Research facilities, both inside and outside of the Chinese automotive industry that could be devoted to automotive research and development are limited in contrast to the facilities available to original equipment manufacturers (OEMs) around the world.
- Even if the physical facilities existed, it remains to be seen whether enough engineers and scientists could be mobilized to carry out the necessary research and development.
- Most of the existing Chinese companies are large, state-owned enterprises that are required to adhere to practices that increase their costs and make them noncompetitive, such as providing housing, schools, and other employee services.

Under terms imposed on members of the WTO, China must discard measures intended to protect an indigenous industry from foreign imports, such as tariffs on vehicles and components. Quotas and import licenses must be eliminated and tariffs reduced by factors of two to four from present levels. These changes, to be phased in over a five-year period, will bring tremendous competitive pressure to bear on China's motor vehicle industry, which enjoys few economies of scale. Some analyses described in Chapter 2 suggest that China's domestic motor vehicle production will have changed little by 2005, but both vehicle imports and auto component exports will grow. The implication is that some of the motor vehicle growth over the five-year period from 2001 to 2005 will result from imported vehicles, signaling that China's motor vehicle industry could face strong competition in its domestic markets during this period.

The future structure of the Chinese automotive industry will likely include a combination of the following forms:

1. stand-alone indigenous OEMs
2. Chinese enterprises, each in partnership in a single distinct joint venture
3. Chinese enterprises, each in partnership in several joint ventures, including some overlap of joint venture members among the various Chinese enterprises, which is common today
4. motorcycle or farm equipment companies capable of developing a small car
5. wholly owned foreign subsidiaries with the capability to manufacture new vehicles
6. small or modest-size entrepreneurial Chinese companies that provide engineering support to all of the various enterprises in China.

Although the first of the alternatives—stand-alone indigenous enterprises—has been assigned a definite role in the five-year plan, it is unlikely that current stand-alone companies will be able to achieve the level of technological independence needed without drastic restructuring. For the next two alternatives, both of which involve some form of a joint venture, it is important to recognize that the technology contained in the world-class cars and trucks currently manufactured in China has been provided largely by the joint ventures with multiple partnerships. But these joint ventures have not provided for the *transfer of the intellectual property* that would have allowed the Chinese members of the joint ventures to develop their own capabilities. With China's accession to the WTO, imported vehicles of higher technical content may displace the Chinese-made cars from the marketplace.

The second form—Chinese enterprises, each in a single joint venture—would have the benefit of allowing a close working relationship to develop between the joint venture partners. Such a relationship could lead to the transfer of knowledge and the building of a technical capability in the Chinese partner.

The third form, a joint venture with multiple partners, is the most common one today, but it generally has not resulted in the transfer of technology.

The fourth form—the entry of either motorcycle or farm equipment manufacturers into the small car market—represents an interesting alternative for China. Many of these companies are freestanding, successful developers of products for their markets. These two industries could become an important part of the restructured automotive industry.

With China's entry into the WTO, the fifth form—wholly owned foreign subsidiaries in China—is likely to appear for the first time in modern history, and even some companies that are presently joint venture partners of Chinese firms may find that this is a more advantageous way to participate in the growing Chinese market.

The sixth form of the industrial complex is the small or modest-size indigenous entrepreneurial company able to provide engineering support to all of the various Chinese enterprises. These independent engineering companies can provide important consulting and development capabilities.

Recommendation

- *A competitive Chinese industry must have access to the results of research on both conventional and advanced technologies if an efficient, low-emission China car is to benefit from technological advancements. Agreements and relationships with joint venture partners should be restructured to allow the Chinese partner to participate more fully in re-*

search and development. Greater participation may require limiting the partnership to one foreign company per Chinese enterprise, with appropriate sharing of technology and knowledge.

TECHNOLOGY RESEARCH AND DEVELOPMENT

Designers of efficient, low-polluting vehicles must treat the entire vehicle as a system, including the engine, transmission, tires, vehicle materials, vehicle aerodynamics, and fuels. Life cycle comparisons of the technological alternatives will be an important tool in ensuring that the proper trade-offs are made. This method takes into account the costs of materials, the manufacturing process, fuels, maintenance, and the disposal of obsolete vehicles.

Conventional power trains have great potential to both improve engine efficiency and reduce emissions. The technology for removing pollutants from the exhaust of spark ignition engines involves sophisticated electronic controls of the power train to ensure the effectiveness of catalysts. The technology for removing pollutants and nitrogen oxides (NO_X) from the exhaust of diesel engines is evolving and will require active management of exhaust system temperature and fuel/air ratios. Over the short to medium term, most energy and emissions improvements are expected to come from dramatically improved conventional power trains.

Hybrid power trains also offer improved fuel economy and reduced emissions. The most efficient hybrid systems are able to turn off the engine while idling and to recover the energy that is available during braking through regeneration. However, cost and complexity remain a barrier to broad implementation.

Onboard energy storage for vehicle propulsion is becoming a more significant issue for some advanced-technology vehicles. Petroleum fuels have become the primary transportation energy source because of their high energy density and simple storage container requirements. Electric vehicles, at the other extreme, are limited in usefulness because of today's heavy and bulky battery technologies. Other alternative fuels are presently at a significant disadvantage because of their added cost or lack of infrastructure.

Worldwide, many companies are investigating fuel cells as alternative power sources for vehicles, and the Chinese government has announced a RMB880 million (over $100 million) fund to support research on fuel cells and other technologies. Although the fuel cell offers the potential advantage of no emissions, the technology is in the early stages of development. Hydrogen, the most promising fuel for the fuel cell, is available from several sources, but the infrastructure to deliver it to vehicles is not in place. Vehicle-mounted reformers, which could use gasoline or methanol and convert them to hydrogen, add weight, complexity, and cost, and reduce overall efficiency and increase emissions. The cost of the fuel cell system, including the cost of the fuel cell and its fuel, must be reduced by at least an order of magnitude before it represents a viable commercial product. In addition, providing the infrastructure to supply hydrogen to a geographically dispersed customer base will cost many hundreds of billions of dollars.

Finally, although the Chinese government has examined the suitability of domestic material resources for manufacture of today's vehicles, similar analyses should be undertaken for the materials that may be needed for the next generation of vehicles. Furthermore, provision should be made for recycling the waste streams generated from scrapped vehicles.

Recommendations

- *Manufacturers around the world are seeking significant improvements in conventional automotive technologies, and Chinese manufacturers risk falling behind if they fail to sustain comparable research efforts on conventional power train systems. The Chinese automotive industry also should strengthen its efforts to develop improved diesel and spark ignition technology in cooperation with its joint venture partners. Researchers should focus on, among other things, advanced gasoline and diesel engine technologies, an ultra-low-emission gasoline engine system, diesel particulate filters, de-NO_x catalysts, selective catalytic reduction (SCR), and improved in-engine combustion management. Industry must develop the capability to model the vehicle power train system in order to optimize its overall performance, including fuel economy, emissions, and vehicle drivability.*
- *Government, in partnership with industry, should continue to support research on emerging and advanced technologies such as hybrids, fuel cells, and battery electric vehicles. Industry can then maintain expertise in state-of-the-art developments and contribute to breakthroughs as they occur. Government, guided by industry, should not single out particular technological solutions prematurely, but should continue to support and encourage the development of a variety of emerging and advanced technologies that are being explored internationally. The long-term, sustainable research and development needed for emerging and advanced technologies requires sustained financial and intellectual support.*
- *So that it can adopt European state-of-the-art emissions standards, which the government has identified as its next level of emissions control, China should adopt fuel quality standards identical to those of the European Union. Achieving these will require that the sulfur levels in fuels be reduced. Because China's present refineries are limited in their capacity to*

produce the cleaner grades of diesel and gasoline fuels that will be required to meet new emissions standards, major upgrades and the construction of new refineries, improved production efficiencies, and expanded foreign partnerships will be required. Substantial investments by the domestic industry, the private sector, and the government will be needed to support the transition of China's petroleum industry into a competitive supplier of transportation fuels.

- *The "China car" envisioned by the government should utilize technologies that allow high fuel efficiency, while providing reliable, comfortable (room for four passengers and some baggage), and safe transportation at an affordable cost. It also should have some attractive attributes that differentiate it from most imported cars in the growing Chinese market.*
- *Although at present there are no fuels that could serve as a practical and competitive large-scale alternative for petroleum fuels over the next decade, research and development on alternative fuels and the associated vehicle technologies, particularly those for which China has a comparative advantage such as hydrogen and liquid fuels from coal, is a wise investment for the future. Serious research in this area will allow the Chinese automotive industry to stay abreast of worldwide developments and perhaps develop new energy sources.*

URBAN ISSUES

Urban residents, with their higher-than-average incomes, are able to buy automobiles earlier and at a higher rate than the general population. As a result, rapid motorization is likely to produce its earliest and severest problems in cities. Because urban vehicle fleets grow more rapidly than urban road length, city-wide congestion increases quickly, leading to some adjustments in urban land use as urban populations move farther from city centers. These patterns are already observable in China's largest cities.

Chinese cities have an especially delicate balance of space use. The experience of cities that have undergone rapid motorization in the past reveals that a growing vehicle fleet results in lower average transit velocities, longer commuting times, and larger areas affected by congestion. As described in Chapter 6, Chinese cities have very high residential densities, and a low proportion of space is dedicated to streets, in accord with the urban designs of two decades ago when bicycles were expected to dominate transport. As a result, these cities are subject to new social forces as increasing congestion, pollution, and the desire for more open space drive people to seek employment and housing at the urban periphery, newly accessible by increased motorization. The net effect of these forces, for better or worse, will be decentralization of employment and residence in the larger cities.

Cities in other countries have responded to these challenges by constructing new urban highways, projects sometimes undertaken by concession to private firms and financed by tolls. When the value of land for parking becomes competitive with that for other uses, investors will seek property for use as private parking spaces near city centers. Municipalities also will build additional public transit facilities, principally for people still without cars in increasingly auto-oriented cities. Worldwide experience shows that few trip makers will change from auto to public transportation in the short run, but employers will increasingly provide private bus services to get employees to work. Eventually the cities will reach new equilibria within expanded geographical areas.

Over the short and medium term in China, automobiles will be available only to a minority of the population because of cost, even in urban areas. But the adverse effects of rapid motorization—such as the increased congestion in the cities which will reduce the performance of public transport—will affect the nonmotorized majority as well. Bicycle use will become less convenient and more dangerous, and exposure to air pollution will rise. These effects may become an equity issue with political or social consequences. A debate between those attempting to accommodate motor vehicle traffic and those defending bicycles as effective and environmentally benign transport for lower-income people has already begun in China.

To relieve congestion and limit pollution, some cities have adopted additional local restrictions on the use of two-wheeled motor vehicles and trucks, in some cases banning them altogether at some parts of the day. A wider concern for the efficiency of the economy will likely engage the central government in these issues.

Recommendations

- *To benefit from the advantages of motorization and avoid its potential problems, municipalities must initiate or expand comprehensive development planning that would embrace transportation, urban land use, and urban services. Such planning should discourage excessive decentralization and promote orderly urban growth.*
- *China's cities should provide additional road space and improve traffic management, while minimizing social disruption to the extent possible. The new construction should not be limited to ring roads and flyover highways; arterial roads also should be improved to relieve congestion in the neighborhoods and business areas, while increasing the proportion of urban space dedicated to transportation. Meanwhile, the available mechanisms for road pricing and other forms of traffic demand management should be explored to provide incentives for more efficient use of road*

space and increased financial resources for road maintenance and construction. Systematic attention to parking requirements and safety will be required as well.
- *As motorization proceeds rapidly in China, it is imperative that more attention be directed to providing public transportation that is convenient, comfortable, comprehensive, safe, and inexpensive. China must create a balance among public transportation, nonmotorized vehicles, and private cars to ensure that the nondriving majority, including bicyclists, is adequately served.*

ENVIRONMENTAL AND HEALTH ISSUES

Increasingly stringent regulations are being imposed on emissions in all developed countries. Although China has not imposed the severest standards, it is following the lead of other countries in requiring automakers to reduce polluting emissions (see Chapter 7).

China is anticipating that its total vehicle fleet, not including motorcycles, will increase by three to seven times between 2002 and 2020. The number of automobiles in the fleet is expected to increase by three to nine times in the same time period. Estimates of the growth of emissions and fuel consumption between 2002 and 2020 based on the midlevel growth scenario described in Chapter 2 (8 percent annual growth of GDP) indicate that motor vehicle emissions of carbon dioxide (CO_2) will more than triple, of carbon monoxide (CO) and hydrocarbons (HC) will almost triple, and of nitrogen oxide (NO_x) and particulate levels will stay at their currently high levels. If the highest-growth scenario (10 percent annual growth of GDP) should occur, even with the current emissions standards (European Emission Standard II, or Euro II, by 2004) and with the 10 percent improvement in vehicle fuel economy specified in the five-year plan, emissions of all pollutants will increase. China should therefore continue to limit vehicle emissions. If growth can be constrained to the medium case, and if emissions standards are aligned with those of the European Union as planned by 2010, vehicle emissions of total hydrocarbons (THC), carbon monoxide, nitrogen oxides, and particulate matter may eventually fall. An aggressive vehicle fuel efficiency program, possibly including an increase in the proportion of diesel-powered vehicles, also will be required to slow the growth of CO_2 emissions.

Recommendations

- *If China is to mitigate its hazardous air pollution problem and the serious health consequences, the Chinese government will need to adopt more stringent vehicle emissions standards nationwide. China should align it-*

self with the European Union's new vehicle emissions standards as quickly as possible but no later than 2010. Meeting future emission standards will require a reduction in sulfur levels in vehicle fuels to below 2001 levels, in line with European requirements.

THE ROLE OF GOVERNMENT

In China, increased motorization, and the regulation and services it requires, will strain government at all levels. The various agencies and institutions that deal with the environment, public health, traffic management and policing, trade, licensing, and regulation will need to be appropriately staffed and funded for their increased responsibilities.

In the United States, Europe, and other industrialized countries, governments have pursued a variety of means to advance technological change in the automotive industry and alleviate negative side effects. Sometimes they have prescribed specific technologies, such as seat belts and other safety features, but the preferred and most effective tool in reducing the adverse impacts of vehicles on the environment has generally been performance standards that allow industry to devise innovative ways to comply.

Governments use a variety of fiscal instruments, in addition to command and control regulation, to influence the choice of vehicle attributes such as low emissions and good fuel economy:

- Vehicle taxes have been used to achieve various objectives, ranging from promoting one technology over another (e.g., diesel versus gasoline cars), to encouraging consumers to select cleaner over dirtier vehicles and more efficient over less efficient vehicles.
- Fuel taxes also can be used to encourage consumers to select more efficient over less efficient vehicles or diesel over gasoline.
- Direct public investment (DPI) can assist industry in developing technologies with certain attributes. One example is the U.S. Partnership for a New Generation of Vehicles (PNGV) program in which government joined industry in funding research. DPI also may be used to stimulate investment in research and development and to support the appropriate industrial structure.

It is observed that in many areas China has put in place regulations (e.g., on emissions, fuel quality, and crash-worthiness) that are quite advanced and draw on the experience of other countries.

The principle of decentralization of power and responsibility, with assignment of responsibilities to levels of management or government where the relevant knowledge resides, has been used effectively for control and regulation in the United States and Europe. Under this principle, the national government retains responsibility for performance or technology standards for the entire country in order to maintain a single domestic vehicle market. While doing so, it takes into account externalities such as greenhouse gas production, national energy security, macroeconomic management and trade, and market failures (e.g., a shortage of available insurance protection). Provincial and local governments retain responsibility for vehicle use and traffic management. In China, this division of responsibility between national and local authorities would enable manufacturers to design vehicles for sale throughout the country and benefit from economies of scale, while enabling local officials to determine the best management of vehicle use to protect local environments and limit congestion.

To increase the technological options that comply with clean air and efficiency regulations, the U.S. and European governments have entered into partnerships with industry for the development of advanced automotive technologies. The U.S. PNGV program illustrates the strengths and weaknesses of the approach (see Chapter 8). The successes generated by PNGV depended on leveraging ongoing government and industrial R&D programs, especially the capabilities of U.S. government national laboratories, and limiting the collaboration to precommercial technologies. Nevertheless, most of the results of industrial research remained proprietary. The benefits of the program to the country were enhanced by the challenge presented to nonparticipating companies such as Honda and Toyota, which advanced their own technology development and put efficient hybrid cars on the market ahead of the U.S. manufacturers. In January 2002 the U.S. government announced its intention to cancel the program and replace it with one directed at fuel cells and hydrogen fuel.

For the Chinese government, its focus on achieving automotive technological improvements may become irrelevant unless high-quality fuel is available on a national scale. In this transitional period, China has competing national and local fuel specification standards. Such competition introduces production complications and inefficiencies for industry that may be mitigated with national standards that are consistent with global standards. Variability in fuels to meet local climatic conditions both seasonally and regionally can be managed within the context of national standards.

Recommendations

- *In view of the very rapid growth forecasted for China's vehicle fleet for the next two decades, environmental and fuel consumption patterns in China could face severe stress, with serious public health and economic conse-*

quences, unless vehicle technology is substantially upgraded. National performance standards have been used successfully worldwide as a means of requiring important vehicle attributes such as low emissions, fuel efficiency, and safety. An appropriate mix of performance standards and incentives may enable China to advance from some of the dated technology in use today to the global state of the art in time to mitigate these problems. Furthermore, the Chinese government and the Chinese automotive industry should develop and implement a process that will regularly assess, nationwide, the appropriate levels of vehicle performance standards, fuel standards, vehicle fuel economy standards, and infrastructure. The government could provide incentives to industry and consumers to accelerate progress in meeting emissions and fuel economy targets.

- *To meet the need for new technologies and greater environmental protection, the government should organize and support government laboratories and academic institutions so that they can pursue jointly with industry research and development on new-generation technologies at the precompetitive stage, with the objective of applying innovations to commercial products. To ensure that the necessary capability and competence exists in the automotive industry, the government should support training of technical personnel at all levels. Smaller companies should not be excluded from the process.*

While this study did not attempt to analyze the full economic consequences of the various actions that will accompany the move toward motorization, the funding requirements will be large. Private expenditures on vehicles and their operating costs will obviously grow. As noted in earlier in this report, both national and local governments will be called on to support major projects. Based on the experience of other countries, annual investments in infrastructure for motorized vehicles are likely to absorb from 1 to 2 percent of GDP and from 5 to 10 percent of public investment—with funds coming from central, provincial, and municipal governments (World Bank, 1994). In addition to road infrastructure, the expansion of joint R&D programs with industry, and of engineering research and development at major universities, will require substantial resources. The committee encourages careful examination of the funding mechanisms that China will use to accomplish its objectives. While endorsing the principle that user fees should be used to finance road facilities, for example through vehicle and fuel taxes, the committee is not prepared to make detailed recommendations about how the infrastructure, technology, and research expenditures associated with motorization should be financed. Because the resource cost of motorization will be large, the analysis of its appropriate funding will require a review of public revenue and expenditure practices at various levels of government.

REFERENCE

World Bank. 1994. World Development Report: Infrastructure for Development. New York: Oxford University Press.

Appendixes

Appendix A

The Development of Personal Use Vehicles for China in the 21st Century

MEMORANDUM OF UNDERSTANDING

This Memorandum of Understanding is hereby made between the Chinese Academy of Engineering and the National Research Council through friendly negotiation in a bid to jointly promote exchange and cooperation in technological and industrial fields. The parties agree to make every effort to realize a collaboration to carry out the following study.

1. Program objective and framework

The objective of the study is to develop recommendations for future approaches to personal use vehicles that are available to all socioeconomic levels in the 21st century.

The outline of the cooperative study for personal use vehicles prepared jointly by both parties is shown in Annex B-1.

2. Organization and members

The Cooperative Study Program shall set up a Committee of approximately 18 persons jointly by CAE and NRC (hereinafter called Committee). Committee members will be selected by CAE and NRC according to the objectives and needs of the Study. There shall be two Co-Chairmen, one from each side. Each Co-Chairman is responsible for organizing his/her own party, and also coordinating with the other party. Committee members shall work according to the principle of division of labor with individual responsibility. If necessary, the Committee shall invite outside advisers and commission papers and reports to inform the Committee on selected topics. NRC procedures for Committee selection and report review will be followed by the parties.

Staff for the study shall be provided by both parties, and shall work together in the execution of the study and the preparation of the report. Funds for the project shall be secured by each party, as follows:

The travel of U.S. members of the Committee and staff to China and the costs of local travel and lodging and meal expenses of the entire committee and staff when meeting in the United States shall be provided by the NRC. The travel of Chinese members and staff to the U.S. and the local transport and lodging and meal costs of the entire committee and staff when meeting in China shall be paid by the CAE. The costs of all commissioned papers and special reports shall be paid by NRC. The costs of translating all working documents and the final report into Chinese and Chinese documents into English shall be borne by CAE. No NRC funds can be used to pay fees or salary to committee members.

The study will begin as soon as the necessary funds are available, and the Committee will attempt to complete the study in about 18 months.

The report shall be prepared in English and Chinese and be reviewed according to procedures of the NRC. Each party will publish the report in its own language.

This Memorandum of Understanding is signed in Shanghai, China, by the authorized representatives of both parties on October 27, 1999.

Guo Konghui	Harold Forsen
(Chinese Academy of Engineering)	(National Research Council, USA)

ANNEX A-1

The Development of Personal Use Vehicles for China in the 21st Century

Statement of Task
(Instructions to the Committee)

The Committee on the Future of Personal Transport Vehicles in China, appointed by the U.S. National Research Council and the Chinese Academy of Engineering, will prepare a report with the following objectives.

1. Review the present status of personal transport in China, the United States, and other countries, including present technology, regulation, and infrastructure in China and the U.S., and problems and achievements in other countries.
2. Analyze the predicted demand for private vehicles in China in the 21st Century.
3. Identify issues that China should explore in developing an automobile industry, including the economics and competition in the international auto industry, infrastructure requirements, congestion, pollution, and sprawl, and experience of the U.S. Clean Air and other relevant environmental legislation.
4. Present and discuss the technological options for Chinese vehicles in the 21st Century. Draw on experience of the U.S. Partnership for a New Generation of Vehicles (PNGV). Consider electric, hybrid, fuel cells, smart cars, and other technologies. Include the roles of alternative modes of transport, public and private, and two-wheeled vehicles.
5. Describe the characteristics of a Chinese New Generation Vehicle (CNGV) in terms of energy efficiency, performance, and manufacturing requirements. Outline a strategy for development of the CNGV, including possible arrangements for government-industry and international collaboration partnering and joint development. Discuss the effect of new generation vehicle production on sustainable development, and economic, social, and industrial structures.
6. Review model approaches to urban transportation and land use planning, including consideration of institutional capabilities for controlling land use. This could include examination of the effects of a CNGV on a selected Chinese city.
7. Formulate recommendations, research directions, and policy choices for China in the 21st Century.

ANNEX A-2

Committee members to include the following skills:

1. Automotive engineering for industry, with knowledge of PNGV
2. Automotive energy alternatives for advanced vehicles
3. Transportation infrastructure
4. Urban planning
5. Economics
6. Forecasting models
7. Technology of internal combustion engines, fuel cells, batteries, fuels, and emissions controls.

In addition there will be one Chinese and one U.S. Co-Chairman.

Appendix B

Case Study: Shanghai, China

Daniel Sperling, Institute of Transportation Studies, University of California, Davis
Lu Ximing, Shanghai City Comprehensive Transportation Planning Institute
Zhou Hongchang, Tongji University, Shanghai

Shanghai is experiencing rapid economic growth. Affluence is motivating dramatic and far-ranging changes in urban structure, transportation, and energy use. This report examines two transportation trajectories that Shanghai might follow.

Shanghai's metropolitan population of about 16 million people[1] continues to grow relatively slowly, but its economy is growing rapidly. The average per capita income is roughly $4,000,[2] three times higher than that of the rest of China, and the Shanghai economy is expected to grow at more than 7 percent a year through 2020.

Massive new transport system investments planned for the next two decades are aimed at lowering Shanghai's extremely high population density, supporting economic growth, and enhancing the quality of life. The list of new investments is impressive: expansion of the new airport; construction of a deep-water harbor, three new bridges, and tunnel river crossings; completion of a 200-kilometer (km) modern rapid transit rail system; expansion of suburban highways; and construction of 2,000 km of

[1] There is considerable disagreement about Shanghai's population. According to the official statistics, the Shanghai metropolitan area (including some rural areas) has 13 million registered residents, but it is estimated that 3 million more people also reside there.

[2] City-level income data are scarce and highly unreliable. One household survey found per capita income in 2000 to be RMB11,718 (State Statistical Bureau, 2001:311) or $1,415 at the official exchange rate (RMB8.28 = U.S.$1). Other estimates measured as "purchasing power parity" or "gross city product" are three times or more greater (see State Statistical Bureau, 1998).

new and upgraded urban roads. These investments will improve the city's transportation system, but are costly and threaten greater energy use and air pollution.

A central issue in Shanghai's development is the role of personal vehicles, especially cars. The city currently devotes little land to roads and has only 650,000 cars and trucks, very few of which are privately owned, placing vehicle ownership levels well below those of virtually all cities of similar income. Even with this small number of vehicles, Shanghai already suffers from serious transport-induced air pollution and traffic congestion.

Shanghai city planners project a quadrupling of cars and trucks in the city by 2020. This projected increase is premised principally on two factors: (1) rapid income growth, which will make car ownership possible for a much larger segment of the population; and (2) falling vehicle prices resulting from China's imminent accession to the World Trade Organization (WTO). Prices are expected to fall because of increased competition, compulsory reductions in vehicle tariffs, and easier access to consumer credit.

The magnitude of the increase in vehicle use is not certain, however. Even apart from WTO membership, vehicle ownership and use—and their environmental implications—will be strongly influenced by three interrelated policy debates: industrial policy toward the automotive industry, air quality policy, and transportation and urban growth policy.

The city's decisions about vehicle use will be critical in shaping Shanghai's future. In this case study, which addresses the forces about to transform Shanghai's transportation system, two transportation scenarios of the future are constructed, drawing upon extensive interviews with decision makers and experts in Shanghai and Beijing. One scenario is premised on rapid motorization, the other on dramatic interventions to restrain car use and energy consumption. Neither is a "business-as-usual" scenario, because this characterization is meaningless in a time of massive investments and policy shifts. Rather, these scenarios are meant to characterize two competing transportation trajectories, taking as given the projected strong economic growth. If the economy grows more slowly, motorization will be slower. Even in the most conservative scenario, though, vehicle travel, vehicle ownership, and energy use increase dramatically.

Caution is urged in generalizing the findings of this case study to other cities in developing nations. Shanghai is not a typical Asian city, given its surging economy and its world-class planning capabilities and strong government institutions. However, the conditions for reining in growth are more propitious here than perhaps any other megacity of the world. If the city is effective at restraining vehicle use, Shanghai may serve as a model for other cities in the developing world.

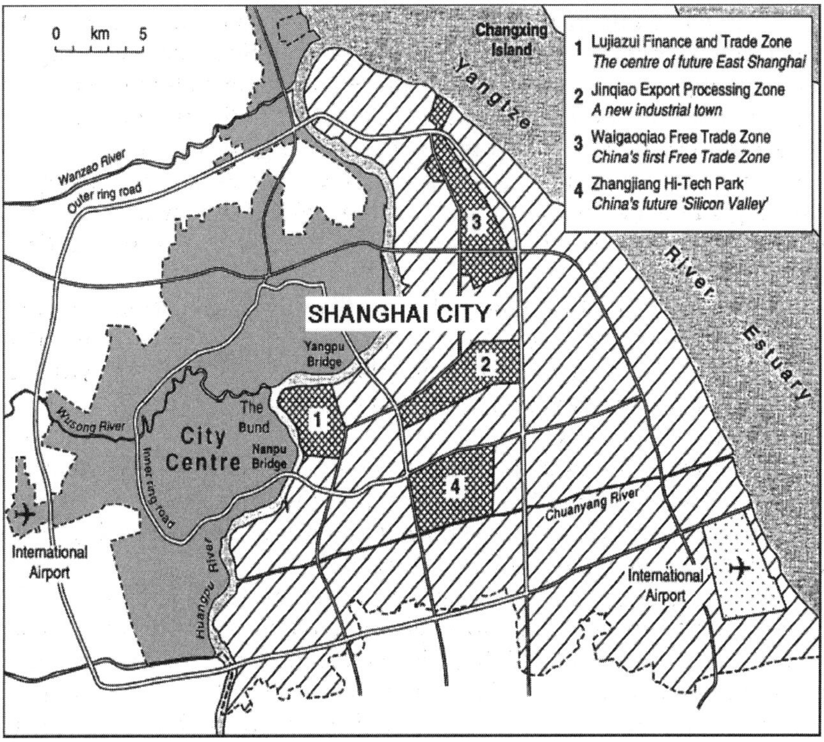

FIGURE B-1 Shanghai. SOURCE: Re-created from Wu (1999) and Shanghai Map Publishing House, 1997.

SHANGHAI: A CITY IN TRANSITION

Sixteen million people reside in the 6,340 km^2 of Shanghai, located on the eastern coast of China in the Yangtze River Delta. The population density of the central city currently averages 22,700 persons per square kilometer. The densest area exceeds 60,000 persons per square kilometer, roughly three times that of Manhattan (Mei et al., 1998:126; Kenworthy and Laube, 1999:429; Wu, 1999:210).

Much of the total land area is rural (Mei et al., 1998:119). The older urban area comprises 280 square kilometers, and a newly urbanized area on the opposite side of the Huangpu River covers another 130 km^2 (see Figure B-1). The urban area of Shanghai is thus about twice the size of Washington, D.C. (Kenworthy and Laube, 1999:393). As a result of market forces and deliberate planning policies, city authorities expect the urban area to expand from 410 km^2 today to 1,100 km^2 in 2010 (Mei et al. 1998:128; Wu, 1999:210, Figure 2).

Shanghai is one of only four cities in China to have the status of a province rather than a municipality. As a result, Shanghai has a higher profile and greater access to national funds than most other cities. Even so, infrastructure spending in Shanghai was low until the 1990s. Because of historical political considerations, the central government did not return a proportionate share of the large tax revenues collected in Shanghai. Housing was in bad repair, as was commercial and industrial space, and road capacity per capita was among the lowest in the world (Wu, 1999:209).

These conditions have changed. Infrastructure funding from local and central government sources, domestic and foreign investment, and international loans have sharply increased (Wu, 1999:207). Massive construction of office and residential space, transportation infrastructure, and public utilities is under way.

This massive investment in infrastructure is due partly to the city's thriving economy. The city has grown faster than the national average, and is widely expected to exceed the nation's forecasted economic growth of 7 percent a year into the foreseeable future.[3]

A central feature of Shanghai's development plans is to reduce its high population density. The local planning authority is pursuing a plan of multicentralization by building eleven satellite cities to siphon portions of the population away from the dense core. Substantial relocation of industry to these cities has already occurred, and many high-rise apartment buildings are under construction. Multicentralization is not a unique phenomenon or goal; it is the de facto or formal planning strategy of most major cities around the world, though Shanghai is pursing this goal more aggressively and deliberately than most.

SHANGHAI'S TRANSPORTATION PICTURE

Despite rapid economic growth, vehicle ownership remains remarkably low in Shanghai. Meanwhile, the city has been investing huge sums in road and rail infrastructure, in part to support decentralization of the city. More infrastructure, satellite cities, and population dispersion will mean more cars, energy use, and environmental stress (see Box B-1).

Shanghai's development has been shaped by its historical role as China's largest seaport. Railways, highways, inland canals, and ocean ship-

[3] See State Statistical Bureau (1998) for Shanghai growth rates. For forecasts of future national growth, see Stiglitz (1997).

BOX B-1
Comprehensive Transportation Planning for Shanghai City

Among major Chinese cities confronting rapidly accelerating motorization, Shanghai has the lowest ratio of cars to population. Although Shanghai is not free of pollution or congestion, it has less of each than Beijing and Guangzhou. This situation has been achieved largely through the use of regulations, incentives, and fees imposed by Shanghai's municipal government to preserve the city's unique character and environment. Under the five-year plan for the automotive industry, however, policy related to automobile ownership and use will be coordinated at the national level. Shanghai is therefore in the process of adjusting its planning so that it can join the national movement toward motorization while protecting commerce, transport, and the urban environment. Shanghai's experience with transportation management and the options under consideration may be useful to other municipalities facing similar challenges.

Those assessing the future economic development of China, including the Shanghai region, predict a 10 percent increase in the gross domestic product (GDP) during 2001–2005, the years covered by the tenth five-year plan. The increase in Shanghai is uncertain, with growth during the next 15 years projected to be about 7 percent. The city's population is estimated to reach from 17.5 to 21 million and the number of motor vehicles to reach between 2 and 3.5 million. For Shanghai, the implications of the expected population and transportation pressure are clear: planning for the suitable development of transportation facilities must begin immediately.

For planning purposes, Shanghai has adopted three principles: development, integration, and prioritization. The development principle means attempting to achieve moderate growth, while preparing for higher growth and avoiding wasted resources if the growth should be lower. Planners must dynamically balance the improvement in transportation services with the rate of growth, try to develop transportation facilities moderately ahead of schedule, and adopt an effective system for traffic demand management. The integration principle means integrating all traffic systems into a single organic system, including transportation facilities and land use and economic policies. The prioritization principle calls for investing first in the projects that will have the most important impact, making best use of advanced transportation management methods to create a highly efficient and fair traffic system.

Present Status of Comprehensive Planning

In 1992 a consortium of municipal organizations in Shanghai completed the Shanghai Comprehensive Transportation Planning system, SCTP1, with the technical assistance of overseas experts. Since then, the population and the state of motorization have changed as a result of economic development policies. At the end of 2000 a revised plan, SCTP2, was announced, based on the second citywide transportation research survey in 1995 and a series of other commissioned studies.

The studies noted a series of specific problems with the current transportation system that require correction:

- The system lacks integration among the different travel modes within the public transportation system.
- The capacity of roads and the coverage of the rail network are insufficient.
- The level of management and service is still low. Because the roads are crowded, bus schedules lag and compete ineffectually with bicycles and motorcycles. Some roads are underused, and the rail transportation system is not used efficiently.
- The traffic flow and environmental quality are not good. Pedestrians, bicycles, and autos are jammed together, resulting in high accident rates and worsening pollution, especially from motorcycles.

SCTP2 will attempt to prepare Shanghai to meet the future challenges just described, and, in doing so, will adopt a focus that extends beyond the city center to the entire metropolitan area.

During the interval between the completion of SCTP1 in 1992 and the initiation of SCTP2 in 2000, the city transportation system did not stand still. During the 1990s the length of roads increased 40 percent, to 6,829 km, the reach of public transportation increased 13 percent, to 23,007 km, and the total number of motor vehicles increased 250 percent, to nearly 700,000. Finally, three new subway lines were added, with a total length of 65 km.

Furthermore during this period, the standard of transportation service was upgraded significantly, and public transportation became more diversified. Taxis made 2 million person-trips a day. During rush hour, the full loading rate of buses was reduced from 8–9 persons per square meter to 5 persons per square meter. The road system also was improved through the addition of 65 km of overhead freeway and the widening of arteries and main streets, and an Adaptive Signal Timing System was adopted to keep traffic running smoothly. Meanwhile, during the 1990s more than 1 million residents moved from the city center to the periphery and the suburbs.

In all these changes, the general goal was, and still is, to construct an accessible, convenient, efficient, safe, reliable, and low-polluting transportation system that is up to international standards and conforms to Shanghai's particular characteristics. Shanghai's comprehensive transportation system will consist of four linked elements. The *passenger transport system* will be based on public ground transportation, with taxis and ferries as supplements. The *road system* will be based on a framework of freeways and artery roads with evenly distributed local streets, including adequate and convenient parking facilities. The *linking system* with the outside world will include airports, a deep harbor, an information depot, freeways, and high-speed railways, linked to the citywide road system, the passenger transit system, and the freight system. It will have intermodal terminal facilities, with the capability to support the expected passenger and freight traffic. Finally, the *transportation management system* will use advanced technologies to ensure smooth operations, safety, environmental protection, and high efficiency.

The passenger transport system will embrace four distinct public transport services. The rail system will be expanded, with a capacity ratio of rail transportation to buses of 6:4. The rail system will have three levels: citywide freeway, townwide artery, and interborough main streets. Traditional public ground transportation will support more than half of the passenger trips, serving short- and medium-distance passengers and those traveling to areas not covered by rail. Within the public ground transportation system, priority will be given to buses for parking, traffic flow, and passenger transfer nodes. To help limit congestion, the number of taxis will be controlled to reduce the vacancy rate from 50 percent to 30 percent. The role of ferries also will be reduced, with an emphasis on providing more service for bicycles. Finally, terminals will be built to facilitate passenger use of the multimodal system.

The road system will be designed specifically to increase the capacity of the downtown street area. Downtown roads will be classified as freeways, arteries, main streets, or local streets. New, outgoing arteries from downtown will serve the new suburban cities, airports, and industrial areas, with speed limits higher than on ring roads and internodal connectors, for both passengers and freight. Part of the road system will be designated for freight to expedite commercial activity without causing excess congestion of central areas. Bicycle lanes will be constructed, and separation of motor vehicles and nonmotorized vehicles will be maintained. Similarly, the pedestrian environment will be protected, with walk signals and pedestrian malls in commercial areas. A new comprehensive parking system, with fees and space designed to limit auto traffic in the city center, will include public parking lots for the transportation nodes in the suburbs.

Perhaps most important, a traffic management system will be developed to manage the time distribution and space distribution of traffic flow, using methods such as land use management, toll fees, parking restrictions, information guidance for drivers, and restricted area policies. The goal will be to create a modern traffic environment suitable for an international metropolis. The Adaptive Signal Timing System will be expanded and improved. A major feature of the new system will be an Intelligent Transportation System (ITS) based on information technology. The main information resources of the ITS will include real-time traffic flow, socioeconomic information, parking availability, vehicular traffic, freight traffic, police status, and a basic geographic information system. The ITS will enable the Shanghai authorities to monitor and respond to changes in the vehicle population and patterns of use, to employ new roads and other facilities rapidly after they are placed in service, and to evaluate continually the effectiveness of the transportation management system to provide optimal service at all times.

Safety will be a primary goal of the traffic management system, and safeguards for pedestrians and bicycles will receive high priority. Among the measures being considered are designating exclusive lanes for buses in the downtown area, controlling the emissions and noise of motor vehicles, separating motor vehicles from nonmotorized vehicles and pedestrians from vehicles, optimizing signal time slots to reduce the emissions caused by deceleration and low speed, reducing the traffic accident frequency, strengthening inspection requirements for vehicles and roads, and accelerating the replacement of old, poor, and damaged cars to improve the overall standard of Shanghai's road transportation system.

—Lu Ximing
Shanghai City Comprehensive Transportation Planning Institute

ping lines meet in Shanghai to exchange freight and passengers. Since the late 1970s economic activity and intercity movement of passengers and goods have sharply increased. Shanghai's port handles 18 percent of the nation's exports, and ranks sixth in the world in capacity (*China Mingbao News*, September 21, 2000). With the booming economy, the seaport is becoming busier. Land delivery of goods through Shanghai's urban transport system also is increasing. Like almost everywhere else in the world, highway transport of passenger and freight has increased faster than railway and sea transport, and airline transport has increased fastest of all. Thus both passenger and freight transport in Shanghai have gradually shifted to more energy-intensive modes (State Statistical Bureau, 1998).

Intracity travel, on the other hand, has relied on modes of travel that consume very little energy. Until about 1990 almost all travel was by foot, bicycle, or bus. Cars, scooters and motorcycles were rare.

Over the last two decades, bicycles have gradually assumed a larger role, replacing walking, and buses have continued to account for a large share of passenger travel. By the end of the 1980s Shanghai reportedly had the largest urban bus system in the world, and the number of riders was still increasing. But limited funding was leading to lagging investments in network expansion, bus amenities, and service frequency. As a result this deterioration of service, combined with higher personal incomes, other, more personalized modes became relatively more attractive.

Shanghai responded in the 1990s in several ways. To restrain large and growing bus subsidies, it introduced competition into the bus supply system. Other cities in China did the same, but Shanghai pursued change more aggressively than most. In Shanghai the municipal bus company was deregulated, and several independent operating companies were created to compete for operating concessions. Bus data from different sources conflict, but all agree that Shanghai continued to have the largest bus system in China through the 1990s, though passenger volumes were shrinking.[4] Shanghai planners anticipate renewed growth in bus travel in the coming decades, with ridership doubling by 2020. They expect that continuing reforms will strengthen the bus industry and that new road infrastructure will be built to serve buses. Plans include building six elevated busways to facilitate bus travel in congested areas.

Planners expect the doubling of bus ridership in part because of a large overall increase in passenger travel. Residents started traveling more and further in the 1980s and increasingly so in the 1990s not only because of income growth, but also because of industry relocation. The movement of factories from the central city to the periphery created long commutes for many workers. Because the newly developed areas were not densely populated, and therefore not profitable to serve, bus companies provided limited service. And because the commuting distance was often too far for bicycles, motorized two-wheelers (scooters and small motorcycles) became a popular mode of travel.

The automobile population in Shanghai is well below the world average for cities of similar income levels. The vehicle population began to expand rapidly in the 1990s, increasing from 300,000 to 600,000 between 1990 and 1998 (Xia and Lu, 1999:22) and reaching about 650,000 in 2000. Businesses and governments own most of these vehicles. About 40,000 are taxis. Individuals own only about 15,000–50,000.[5] The city government controlled new vehicle registrations with a high vehicle registration fee through 1998. The city has used an auction system for vehicle registrations since then.

---

[4] See, for example, Stares and Zhi (1995b:489) and Chang (1999/2000).

[5] Official sources from 1998 indicate 10,000 privately owned vehicles (see Shanghai City Comprehensive Transportation Planning Institute, 1998; Rao, 1999). Informally, city officials indicate the number is closer to 20,000. But the car manufacturing companies in Shanghai indicate that their employees own about 10,000 vehicles by themselves. Executives at the Shanghai Automotive Industry Corporation, the Chinese holding company for joint ventures with Volkswagen, General Motors, and others, indicate that an additional 30,000 or so employees of major Shanghai companies own their own vehicles but have registered their vehicles through their employers—and thus the city does not record those 30,000 or so as privately owned.

Even with the small vehicle population, the streets are congested—the result of high population density, many pedestrians and bicycles, and limited road infrastructure. Bicycling and walking are the primary means of travel, together accounting for over 60 percent of total trips taken in 1995 in Shanghai (Shanghai City Comprehensive Transportation Planning Institute, 1997a:5). Shanghai residents own 6–7 million bicycles (roughly one for every two residents), plus 250,000 scooters and small motorcycles, and about 500,000 mopeds (less than 50 cubic centimeters).[6] The scooter and motorcycle population is declining in the central city area because of new restrictions on the registration of new scooters and other vehicles with two-stroke engines. These restrictions are premised on air pollution and safety concerns. This decline may be temporary, however. As incomes increase, travel patterns disperse, and cleaner-burning four-stroke engines (and perhaps battery-powered two-wheelers) become available, sales of motorcycles and scooters are likely to surge.

Because most walking and bicycling trips are short, measuring the modal split by passenger-kilometers traveled paints a different picture than measuring it by number of trips, as indicated by Figure B-2.[7] Motorized travel now accounts for about two-thirds of all passenger-kilometers traveled. About two-fifths of that motorized travel is by car and motorized two-wheelers.[8]

Although the absolute number of vehicles is still relatively small, traffic congestion and air pollution are becoming severe. By 1993 transportation accounted for most of Shanghai's urban air pollution, contributing an estimated 90 percent of carbon monoxide, 92 percent of volatile organic gases, and 23 percent of nitrogen oxide (NO_x) emissions. In 1996 monitor-

[6] Written briefing provided by Shanghai government planners for visiting National Academy of Sciences committee, May 16, 2001, and affirmed by senior planners. Also see Ying (1998:155).

[7] Modal shares measured in terms of passenger-kilometers are calculated using estimates of typical travel distances by mode, average loads on each mode, and number of trips by mode. To estimate passenger-kilometers, multiply the number of passengers by the distance traveled. For example, 10 passenger-kilometers could equal 1 passenger traveling 10 kilometers or 10 passengers traveling 1 kilometer.

[8] Not all sources agree on transportation statistics presented here. For example, Shanghai government data indicate somewhat lower nonmotorized shares than World Bank sources (see Chang, 1999/2000:24). These data uncertainties do not, however, undermine the central observation that nonmotorized travel in Shanghai continues to be unusually high, facilitated by the city's high population density and mixed land uses. Mixed land use, meaning combined rather than separate areas for industrial, commercial, and residential use, tends to result in fewer long-distance trips, especially in cities, because shopping, work, and school destinations can be more clustered.

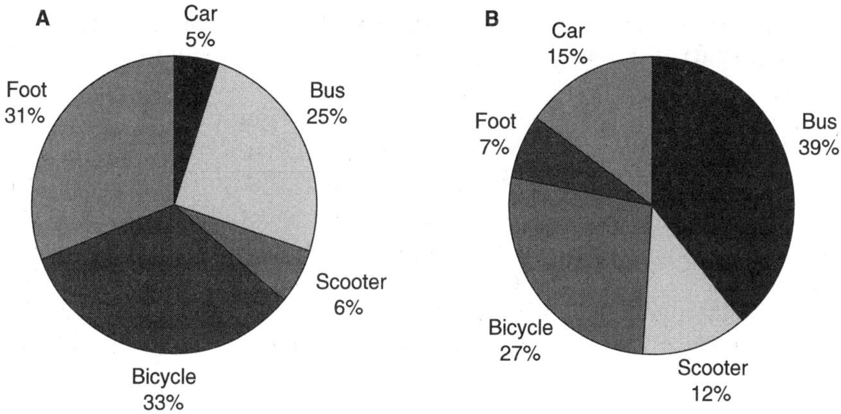

FIGURE B-2 Travel by mode, Shanghai. A. Modal split by number of trips (1995); B. Modal split by number of passenger-kilometers traveled (2000). SOURCES: Lu et al. (1996) and Shanghai City Comprehensive Transportation Planning Institute (1997a).

ing data indicated that transportation accounted for 56 percent of NO_x emissions (Chen, 1997).

To limit air pollution and traffic congestion, city officials began capping the registration of all new cars and trucks in 1998 at 50,000 annually (Shanghai City Comprehensive Transportation Planning Institute, 1998). The government also limits ownership of motorized two-wheelers. In 1996 Shanghai capped the registration of mopeds (under 50 cc), allowing owners to transfer registrations to new mopeds but not to purchase additional mopeds, and soon after banned the use of all scooters and motorcycles (over 50 cc) from the city center. The only unrestricted motorized vehicles are two-wheelers powered by batteries, but few of these are available.[9]

These motorcycles and scooters are unlike those seen in the United States and most of Western Europe. They are very small, with inexpensive two-stroke engines that are inefficient and highly polluting. The largest ones are almost all less than 150 cc, much smaller than most scooters and motorcycles in the United States.

Restrictions on motorized two-wheelers stem in part from their high emissions and noise. They also are perceived as unsafe because they mix with slower moving bicycles and often are driven aggressively by young

[9] A mandate in Taiwan requiring a growing proportion of zero emissions (battery-powered) two-wheelers suggests that this new technology may become more competitive in the near future.

men. The prevailing view in the government seems to be that these vehicles are part of the early stages of economic development that they will soon pass through—a view based largely on the rise and then near disappearance of scooters and motorcycles in Western Europe.

A more complex problem confronting Shanghai is traffic congestion. Serious congestion is relatively new. The problem is quite different from that of most U.S. cities, mainly because of the large number of bicycles and pedestrians sharing the roadways with cars, motorized two-wheelers, and buses. Even freight movement is sometimes performed by bicycle in Shanghai. There is limited road space because land is intensively used for other purposes, making traffic congestion an endemic problem.

Transport Infrastructure: Plans and Investments

Shanghai has responded to pressure on the urban transport system with massive infrastructure investments. From 1991 to 1998 about 14.6 percent of the GCP was devoted to construction—and a significant percentage of that for transportation, a much higher rate than is typical for developing country megacities; the surface area of paved roads increased by 62 percent (Shanghai City Comprehensive Transportation Planning Institute, 1997c, 1998).[10] In 1993 Shanghai spent three times more money on urban construction and maintenance than any other Chinese city, about half on roads, bridges, and mass transit (Ministry of Construction, 1994).[11] From 1991 to 1996 Shanghai spent approximately RMB83 billion ($10 billion) on transport infrastructure, including two major bridges, a tunnel, an inner ring road, 65 km of elevated freeways, and the first line of its new subway system. Wu Weiping writes: "The pace was something like building the Brooklyn and Manhattan bridges in New York and the Lincoln and Holland tunnels between New York and New Jersey all in five years" (Wu, 1999:209). Shanghai has plans to continue with intensive infrastructure investments, building both additional roadway infrastructure and public transit infrastructure.

The major motivation for this burst of activity was to fill the transport infrastructure deficit resulting from decades of deferred investment. By 2000 the projects from the first plan were completed, and the local gov-

[10] From 1991 to 1998 total investment for urban infrastructure was RMB261.7 billion ($US31.5 billion), accounting for 14.6 percent of Shanghai's gross city product for that period. Of the RMB261.7 billion, RMB8.27 billion ($5.81 billion) was reportedly allocated for urban road transport infrastructure construction, representing 2.6 percent of Shanghai's gross city product, but other local government reports indicate the percentage for roads to be as much as twice as great.

[11] The Ministry of Construction report was cited in Stares and Zhi (1995b:491).

ernment declared that the infrastructure deficit no longer existed (*People's Daily Newspaper*, September 25, 2000).

The second phase of the urban transport planning effort began in 1995.[12] It is aimed at moving housing and industry outside the city center to decentralize the metropolitan region. Shanghai's Land Use Master Plan predicts for 2020 a population increase of 2–3 million, a multicenter metropolis with a strong central business district, a new city center in Pudong New Area on the east side of the Huangpu River, and eleven satellite towns, all linked by an efficient transport network.

The second plan also calls for three Huangpu River crossing facilities, a second runway for the new international airport, a new deep-water harbor for container ships, 200 km of rail (of which 65 km were completed by 2001), six elevated busways, and 650 km of divided highway in suburban areas, of which 520 kilometers will be new (Shanghai Highway Administration, 1999:5–11). Roads serving as part of the intercity network will charge tolls. The new rail system will be largely underground and is forecast to carry 8 million passengers a day by 2020.

Vehicle Ownership in Shanghai

The most striking aspect of Shanghai's transport system is the small number of cars and the rarity of private vehicle ownership. As noted earlier, Shanghai has only about 15,000–50,000 privately owned vehicles. Beijing, with similar income and population, has perhaps 10 times more. Even in terms of the number of total vehicles, Shanghai has fewer than most cities of comparable wealth. Shanghai has several times the income of Delhi, for example, but less than half the number of private vehicles (see Table B-1).

The scarcity of privately owned cars is related to issues of access, cost, ease of use, and quality. First, it is expensive and time-consuming to acquire a driver's license (Ni, 2001). One must enroll in an official driving school at a cost of RMB4150 ($500), a significant expense for the typical Shanghai resident. The course consists of three weeks of classroom sessions, more than a month of behind-the-wheel training, and three separate road tests.

Second, it is very expensive to own and operate a car in Shanghai (Gwilliam, 1995:391). Fuel prices are similar to those in the United States,

---

[12] Information on the second transport plan come from Shanghai City Comprehensive Transportation Planning Institute (1997b), a report prepared for the World Bank; Shanghai Metropolitan Highway Network Planning Agency (1998); and a 1999 interview with Lu Ximing, director of the Shanghai City Comprehensive Transportation Planning Institute.

TABLE B-1 Cars per 1,000 Inhabitants, Seven Cities

City	Population	Cars per 1,000 Inhabitants
Santiago	5,500,000	129[a]
Delhi	13,418,000	63[a]
Shanghai	16,000,000	22[a]
London	6,852,000	340[b]
New York City	7,497,000	230[b]
Tokyo	8,164,000	210[b]
Paris	8,791,000	340[b]

[a] 1998–2000.
[b] 1990.

SOURCES: Authors and Focas (1998).

but parking costs $1–3 per hour in downtown Shanghai. The greatest barrier is purchase price. Based on the current exchange rate (RMB8.28 = U.S.$1), the sale price of a small, domestically produced sedan is about $10,000. The actual price is much higher. A tax of approximately 10 percent and a large local registration fee must be paid at the time of purchase. Until 1998 the registration fee was approximately $20,000 on new cars. Under pressure from the central government, the city discarded the high fees and created a vehicle registration auction similar to the one used in Singapore to limit the number of new vehicles that could be registered per month in the city (Stares and Zhi, 1995a:79). In early 2000 the auctioned registration fee was approximately $2,500.

For imported cars, the cost is even higher because of extremely high tariffs. However, in November 1999 China and the United States signed a WTO accession agreement that cut tariffs of 80–100 percent on imported cars (varying by type and price of vehicle) to 25 percent in 2006.

Cost is a barrier not only because it is high relative to average incomes, but also because consumer credit is not yet widely available in China. This means that a prospective car owner must pay the full amount upfront, an outlay that remains beyond the reach of virtually all families. Yet, in addition to reduced tariffs on imported cars, WTO accession will require opening up the financial services market, which should lead to easier access to consumer credit. The result would be much greater ease in purchasing vehicles. The extent to which consumer credit will in fact become more available remains uncertain, however, as does the extent to which Chinese consumers will embrace buying on credit.

A third deterrent to car ownership is limited road infrastructure and severe traffic congestion. Land use patterns in Shanghai evolved before motorized transport. The city grew in a very densely developed radial pattern, with narrow streets conducive to bicycle use and pedestrians. Services, schools, and jobs are well mixed with housing and within easy bicycling distance for most people. Because trips are generally short and bicycles and public transit both widely available, cars bring little extra value for everyday travel. For intercity travel, options include train, bus, or airplane. Road touring vacations are rare in China.

The fourth explanation for low private vehicle ownership rates is the relatively low quality of vehicles that have been available for sale. The tariffs on imported cars are aimed at protecting the fledgling domestic automotive industry, giving substantial market power to local producers. The result is elevated prices for products that are often technologically outdated. Most vehicles have been produced by joint ventures between major international automakers and local companies, until recently using technology from the 1980s.

Motorization in the Coming Decades

Pressure for increased private auto ownership in Shanghai comes from several sources: income growth, car economics, social status, population growth, and population dispersal.

Car economics, and therefore car sales, are affected by government policies in various ways. One is through national industrial policy. In 1991 China designated automotive manufacturing a pillar industry, initiating a major debate, still under way, over the extent to which the government should promote automotive manufacturing. Shanghai is deeply engaged in this debate because it is a major industrial center and already home to one of the three largest automotive manufacturing companies in China.

In 1991 Chinese leaders established a national goal to produce 1.2 million cars per year by 2000 and 3.5 million per year by 2010, with 90 percent of output sold domestically. The policy encouraged private car ownership, eliminated government control of vehicle purchases, reduced taxes, and allowed the marketplace to determine car prices (ITE-SPC, 1994). However, actual production in 2000 of about 0.6 million fell far short of the 1.2 million goal.

During the late 1990s, nearly every province, including Shanghai, encouraged local investment in the automotive industry. The result was excess capacity and a plethora of small, inefficient companies. Many local governments have given up their earlier ambitions, but several joint ventures between local companies and international automakers have taken off. These companies now have the capability to bring the latest technology and products into China.

Although Shanghai never articulated a clear strategy, policy makers presumed that the auto industry would be a boon to the local economy.

Shanghai has been particularly aggressive and successful in this regard. By 2000 auto-related production accounted for 20 percent of GCP (State Statistical Bureau, 1998). Shanghai is home to the largest automotive company in the country, which includes a joint venture with Volkswagen producing over 200,000 cars and another with General Motors (GM) that began production in 2000 and is building a large manufacturing complex. The GM joint venture also is exploring production of simple, inexpensive "agricultural vehicles" that cost from RMB5,000 ($600) to RMB25,000 ($3,000) and are designed for passenger and goods transport in small cities and rural areas.

Some observers suggest that consumers recently have been deferring their car purchases, partly in anticipation of better and less expensive cars becoming available after China's accession to the World Trade Organization. In any case, dramatic increases in car buying are assured in coming years. In Shanghai some of the barriers that deter private car ownership are already being lifted. WTO membership will result in higher car quality and, barring new taxes, lower car prices. Given the huge population and rapid income growth, foreign automakers and parts suppliers are expected to enter the Chinese market in an aggressive manner. This intensified competition, along with the increased availability of consumer credit, will be a strong force for increased car ownership.

Beyond economics, a second reason to expect increased car ownership in Shanghai is status. The private car is both a personal and a national symbol of status and success. Many Chinese believe that when more people own cars in China, the status of the entire nation is elevated in the eyes of the international community. This view of the car as a status symbol is not unique to China.

A third explanation is population dispersal. Shanghai's multicentralization policy will reduce density in the city center by creating multiple urban centers around the periphery. This decentralization will lead to longer trip distances, reducing the attractiveness of walking and bicycling while enhancing the attractiveness of private vehicles.

A fourth explanation for vehicle growth is population growth. Shanghai is an affluent city and an attractive destination for many Chinese seeking a better life.

Yet Shanghai has powerful local institutions that can effectively manage growth. The extent to which they restrain vehicle ownership and use will depend largely on evolving perceptions about the desirability of motorization and larger national policies. The national government continues to pursue a strategy to make the auto industry a pillar of the national economy. In 2000, in support of this strategy, the central government announced that 238 vehicle fees were being eliminated (Energy Foundation, 2000).

Another important, though often ignored, element of the debate over automotive industry investments is motorized two-wheelers. They are often ignored for three reasons. First, China already has a large two-wheeler manufacturing industry; roughly half of all motorized two-wheelers in the world are manufactured in China. Second, motorized two-wheelers require much less investment and manufacturing and engineering capability than cars. Third, as indicated earlier, many Chinese policy makers see large-scale use of motorized two-wheelers as a relatively brief phenomenon in the development process of the country, much as occurred in Europe.

POLICIES AND STRATEGIES

This section examines current and prospective transportation and environmental strategies.

Air Quality and Energy

In the past, air pollution problems in Shanghai owed their existence to the city's heavy industry. Most of these factories have been relocated to surrounding areas, partly to clean up Shanghai's air. Now, a substantial portion of Shanghai's air quality problem is produced by the transport sector, despite relatively few motorized vehicles in the city.

Indeed, the transport sector has become a focal point for air quality issues in many urban areas in China. The national government recently implemented stringent regulations aimed at limiting air pollution from urban transportation sources. Among the new pollution regulations implemented in China are vehicle emissions standards, mandatory inspection and maintenance programs for vehicles in certain cities, and gasoline quality standards. The new vehicle emissions standards are ambitious, equivalent to the standards that first took effect in Europe in October 1993 (known as European Emission Standard I, or Euro I) and in the United States in the early 1980s. The gasoline quality standards include a nationwide ban on leaded gasoline, effective January 2000, which apparently is being observed and enforced. However, sulfur levels remain high in both gasoline and diesel fuel, impeding the introduction of advanced emissions control technology.

Shanghai's city planners have been environmental leaders in China. Air pollution is severe in Shanghai, though considerably less so than in Beijing. Along with several other large cities, Shanghai began eliminating leaded gasoline ahead of the national government in 1998. In July 1999, again ahead of national requirements, the city promulgated new emissions standards for other pollutants, and in the late 1990s began switching many vehicles to cleaner-burning liquefied petroleum gas (LPG) and compressed natural gas (CNG). The taxi fleet is currently being retrofitted to burn LPG, and the bus fleet is being retrofitted to burn CNG. The first CNG station opened in October 1998.

Although the pollutants that cause deterioration of local air quality are largely different from the gases that contribute to global warming, many, but not all, of the strategies to reduce air pollution also reduce GHG emissions. In any case, air quality initiatives are and will be far more effective than GHG emissions initiatives because air quality tends to be a strongly compelling, popularly embraced goal; GHG reduction is not. Still, initiatives to reduce air pollution build environmental consciousness and strong constituencies that carry over to other environmental goals. Moreover, any strategy that restrains vehicle use will generate large air quality, GHG emissions, and energy benefits.

Some believe that widespread use of alternative fuels such as natural gas may be a leading strategy to help solve China's environmental and energy problems (Weisbord, 1999). Indeed, CNG combustion results in much lower air pollutant emissions than gasoline or diesel combustion, though the effect on greenhouse gases is not as dramatic.[13] CNG also is less costly than gasoline and, by replacing gasoline, provides substantial energy security benefits. China is now importing petroleum, and imports are expected to grow. The natural gas situation is more positive. China has larger domestic supplies of natural gas than petroleum, and Shanghai has access to gas from the East China Sea as well as northwest China. Little natural gas is used today in China, but the country plans to exploit domestic sources and import liquefied natural gas. A pipeline from northwest China to Shanghai is scheduled to be completed in 2007 (*Guangzhou Daily*, February 3, 2000, in Chinese).

An important target of air pollution control efforts in Shanghai, and a large source of the pollution are two-stroke scooters and motorcycles. New registrations for these vehicles have not been granted since 1996, but their population remains high because old registrations can be transferred to new vehicles.

The city recently began to promote electric scooters as an option for residents who want the convenience of a new scooter. Motorized two-wheelers are particularly attractive in Shanghai as a means of personal transport because they are faster than bicycles, affordable for many, and

---

[13] The use of CNG in spark ignition engines in place of gasoline results in about a 20 percent reduction in GHG emissions. However, in diesel engines CNG provides little or no benefit and can even increase emissions. These impacts assume that engines are reoptimized for CNG and emissions are calculated on a full energy cycle basis. If the engines are not reoptimized, as with most retrofits (versus redesigning or remanufacturing the engine), then CNG would produce even more greenhouse gases.

easier to park than cars. Several electric scooter companies have established service and battery exchange networks for their customers to increase convenience (*Shanghai Evening News,* February 10, 2000). Electric scooters provide huge air quality benefits and are far more energy-efficient than existing scooters, though GHG benefits are modest, the result of 75–80 percent of electricity in the country being generated from coal.

In summary, Shanghai has already embraced the goal of improved environmental quality. Indeed, most strategies to reduce adverse environmental impacts of transportation are mutually compatible and consistent. Environmental goals also are an impetus to restrain motorization and introduce electric-drive propulsion technologies that use fuel cells or hybridized combinations of electric motors and small internal combustion engines. These advanced technologies are very clean and energy-efficient. Hybrid-electric technologies reduce energy consumption by up to 50 percent, and fuel cells potentially even more.

Avoiding Gridlock

The key question is: Can Shanghai continue to manage the growing desire for personal vehicles? Shanghai faces a difficult challenge in expanding its transport system. Increasing affluence and falling car prices lead to rapid motorization superimposed on a dense city that has minimal road capacity. Without exceptional investments, management, and policy intervention, the city could quickly become gridlocked.

As noted earlier, serious traffic congestion is a relatively new problem for Shanghai. A variety of policies and investments aimed at meeting Shanghai's transportation challenge are already in place. The vehicle population is controlled using a monthly auction system for new vehicle registrations. Freight movement by truck in the central city is restricted during the daytime when passenger travel demand is highest. Over the last decade 130 km of expressway were built, in large part to support the multicentralization process, with another 520 km planned (Shanghai Highway Administration, 1999). Substantial investments have been and continue to be made in public transit, including a new subway system that opened its third line in 2000.

Also important, city planners are strongly committed to enhancing intermodal connections. Pedestrian and bicycle paths are being built to transit stations, ferry service is designed to carry bicycles, and bus routes are designed as feeders to rail transit line-haul service. There are plans to build 44 large-scale passenger transit terminals that will provide for transfers between buses, rail, and ferry boats, and accommodate taxis, private cars, and bicycles. Specific plans are being made to use high-speed magnetic levitation (maglev) rail, bus, and car to enhance access to the international airport. Shanghai planners are expanding and improving multimodal freight terminals. Transfer terminals are being built and expanded at the harbor, airport, and rail stations, and along the city's ring roads.

Transportation is a central concern in Shanghai. A poorly managed transport system can hamper economic growth by creating costly, long, and erratic connections. Shanghai planners and managers understand that a well-managed system reduces costs and eases access, and they seem committed to improving transport to keep pace with the growing economy.

Bicycle Infrastructure

Bicycles produce no pollution and are an inexpensive form of transport, but they are uncomfortable in bad weather, can be unsafe, and are unsuitable for some people. Bicycles use road space more efficiently than cars, but less efficiently than buses. There is widespread agreement among Shanghai planners and leaders that bicycle use should be limited.

One means of making the best use of existing road space is to separate traffic operating at different speeds. The presence of bicycles greatly slows motorized traffic if they share road space. Sharing the road with motorized traffic also can be quite dangerous. On most streets in Shanghai, bicycles and small scooters are separated from the flow of buses and cars with wide bicycle lanes. These lanes are used heavily, improving safety and lessening traffic congestion. Nevertheless, traffic delays occur where these lanes cross intersections (where turning bicycles and cars disrupt traffic flow). Where bicycle use is light, this would not be a problem. In Shanghai, however, this situation can cause significant delays. During busy times of the day, it is common to see more than 50 bicycles and scooters stopped at a red light. At some intersections in Beijing, separate traffic signals have been installed for the two sets of vehicles, but it is not clear that signals improve traffic flow in these cases. Other strategies under consideration are bicycle-only streets (Wu and Li, 1999:49), and intersection overpasses for bicycles and pedestrians.

Information Technologies for Traffic Management

On congested links, even minor accidents or adverse weather conditions can be highly disruptive. The increasing availability of low-cost information technologies now makes it possible to monitor and manage traffic flow in real time. In the mid-1980s Shanghai began installing advanced traffic coordination systems. Approximately 1,000 intersections are now monitored, and 18 different systems are controlling surface and elevated roads, tunnels, bridges, and subway rail.

Unfortunately, these systems are not efficiently linked, and traffic information is not shared among different systems. The city is planning to correct this situation within the next five years through development and implementation of an integrated traffic coordination system (*Shanghai Evening News*, November 16, 2000). The result should be improved efficiency, especially via timed signal lights and rapid removal of vehicle breakdowns. But with widespread congestion and growing travel demand, more fundamental vehicle restraint strategies must accompany advanced traffic management.

Restraining Use of Full-size Private Cars

To restrain use of full-size private cars, policy makers must focus on car purchases. The fixed costs of using a private car for transport are much higher than the associated variable costs, such as tolls and parking fees. In fact, once a person owns a car, much of the price of transportation has already been paid. A car owner will often choose to drive even when convenient and inexpensive alternatives exist. The most effective way to avoid this situation is to offer attractive transportation alternatives and raise the variable costs of vehicle use to reflect environmental and other associated costs. Existing policies include stringent and expensive driver licensing and vehicle taxes. Policies under consideration range from car-free zones in especially dense areas to rules on what types of vehicles companies can build and sell.

Shanghai is already planning to charge substantial tolls for highways being built to serve the new satellite cities (although it eliminated bridge and tunnel tolls in 2000 on the premise that they were redundant with high vehicle registration fees). At least one car-free zone is in operation in downtown Shanghai, and most cargo truck travel is banned during the daytime. Parking in downtown Shanghai is expensive, and taxi service is reasonably priced. A relatively inexpensive but effective option to restrain vehicle use is already being pursued in Shanghai—the creation of strategically placed car-free zones during peak periods in areas well served by public transit. Car traffic would be banned (with the possible exception of taxis) in designated areas during peak travel periods. In many parts of the world, this type of policy would be difficult to implement because of large infrastructure capacity and high car ownership levels. Car owners, being the wealthiest and most powerful residents, would use their political and economic power against the proposed policy. However, China's centralized decision-making structure is relatively more resistant to such pressures.

In 1997 a car-free zone was introduced on Nanjing Road, the main shopping street in Shanghai. At first, the zone was car-free only on weekends, but with the recent construction of an underground rail transit station in the area, car-free restrictions were extended to weekdays as well. Shanghai has also implemented a similar policy on freight traffic. Between 7 a.m. and 7 p.m. most heavy freight traffic is banned from the central city. This practice is common in many large Chinese cities (Stares and Zhi, 1995a:82.). With this precedent, local leaders are likely to accept the creation of more extensive car-free areas in Shanghai during peak periods.

Another policy might be to charge high parking fees, with fees highest in the densest areas, coupled with limitations on parking space. This strategy is already being pursued to some degree in Shanghai, where parking fees are extremely high compared with the average income of a Shanghai resident.

Improving Substitutes to Full-size Private Cars

The most important alternative to the private car is public transit. Shanghai has been investing heavily in public transit in recent years, overhauling the bus system, and constructing an ambitious 200 km heavy rail metro system. The city is also building high-rise apartments and bicycle parking lots near new rail transit stations. The convenience of Shanghai public transit is enhanced with an integrated electronic fare collection system. Since 2000 people have been able to ride all modes of public transit in Shanghai, including buses, rail, and ferry boats, with one transit card (Wang, 1999).

Another alternative to the full-size private car is a smaller private vehicle. Many large auto manufacturers around the world are developing and selling very small cars for crowded city use. In Japan, over one-quarter of new vehicle sales have long been minicars (defined as having engine capacity of less than 660 cubic centimeters [cc]). These vehicles are not suited to long-distance or high-speed travel, but they function well for urban use. They are typically about half the size of a conventional sedan. New, inexpensive models are under development.

Scooters and motorcycles also economize on road space while providing many of the benefits of a personal car. Government policies could favor the use of minicars and electric scooters over conventional sedans by providing preferential parking, reducing fees, and relaxing vehicle registration fees.

Another option that would limit cars on the road is car sharing. This new form of car ownership is becoming popular in Switzerland and Germany (Shaheen et al., 1998; Carsharing, 2000; Gakenheimer and DeLisi, 2000). In Shanghai, car sharing could become the main method of accessing full-size vehicles for the general population. In car-sharing organizations in Europe, members pay a yearly fee plus a charge per hour of use and per kilometer of travel. These fees cover all expenses associated with owning a vehicle, including insurance, maintenance, and fuel.

Good traffic management together with Shanghai's planned infrastructure investments and a forward-thinking approach to vehicle policy could make Shanghai's transport system a world-class model.

Leapfrog Technology Opportunities

Leapfrog technologies are advanced technologies that allow developing countries to go beyond what is typically being used in developed countries. Shanghai could leapfrog the pollution problems and other pitfalls that industrialized countries have encountered on their development paths.

The two scenarios in this study present a set of strategies and technologies that include varying quantities of leapfrog technologies. One leapfrog technology gaining considerable attention is fuel cells. Fuel cells provide the promise of very low air pollution and GHG emissions and high energy efficiency. Even more innovative solutions are possible, though not specifically targeted in the scenarios. They include automated bus rapid transit systems in which groups of buses operate on a network of specialized lanes—an enhancement of the elevated busway system already contemplated. These buses could branch off at either end of a line to collect and deliver riders in less dense areas.

Likewise, small cars with small battery packs or fuel cells could operate on an electrified road, perhaps under automated control. The cars would gain power from the roadway, either from conductive or inductive electricity transfer. With automated control, the capacity of the roadway would be very high (because lanes would be narrow and headway distances between vehicles very small). The vehicles would veer off from the automated, powered roadway at the beginning and end of their trip.

These dual-mode car and bus systems might prove highly efficient from an economic and environmental perspective and provide high-quality service. Except for full vehicle automation, these technologies are technically well within reach of current engineering capabilities. They have not been implemented largely because of an array of financing, liability, and institutional issues. In developed cities, the cost and challenge of retrofitting the existing road infrastructure is daunting. In developing cities, where infrastructure is not yet in place, there is an opportunity to design the transportation system to accommodate these technologies right from the beginning. Given China's enormous population and growing wealth, it may be the time and place to begin testing these revolutionary concepts.

SCENARIOS FOR THE FUTURE

Vehicle ownership and use will soar in Shanghai under any plausible scenario. But with more vehicles and travel will come more air pollution, energy use, GHG emissions, and road infrastructure costs. These impacts can be mitigated by restraining vehicle ownership and use. At times, the impacts also can be mitigated independent of one another and sometimes even independent of vehicle use. They are not necessarily locked into a fixed relationship with motorization. The best, but not only, example of targeted strategies is air pollution control.

Enhanced emissions control technology can greatly reduce air pollution from conventional internal combustion engine vehicles. Indeed, China is already embarking on that path with the adoption of Euro II emissions standards. Energy use and GHG emissions are far more difficult to reduce. In the most extreme case, fossil energy use and GHG emissions could be almost completely eliminated—without changing car ownership levels—by substituting nonfossil energy sources. In less extreme scenarios, energy consumption (and therefore greenhouse gases) could be reduced by reducing the size and weight of vehicles and the combustion efficiency of the engines. Vehicles can be large sedans or small minicars, and they can use relatively inefficient conventional internal combustion engines or highly efficient advanced diesel engines and fuel cells. A small, efficient vehicle, for example, would consume as little as one-tenth of the energy consumed by a large, gas-guzzling sport-utility vehicle. And demand for road space can be reduced, not only be restraining vehicle use, but also by downsizing vehicles, managing roadways more efficiently (for example, with advanced traffic management technologies), and creating new car-sharing ownership mechanisms.

Exactly how Shanghai develops will have far-reaching implications for Shanghai's economic, environmental, and social well-being. Here two scenarios of Shanghai's transport future are postulated. Each is motivated by a different set of political, economic, and environmental conditions.

Scenarios are commonly employed to deal with complexity and uncertainty in forecasting. Ideally, the researchers generate relevant information using credible research methods and objectively analyze it by means of alternative scenarios of the future that provide upper and lower bounds on a plausible range of motorization levels and adverse environmental impacts. The scenarios reflect realistic, but often quite contrary, development paths. This approach can provide a useful context for the development of a "no regrets" public policy and business strategy.

To generate scenarios, the authors interviewed Chinese transportation experts and political leaders in Shanghai and Beijing. They also analyzed historical data and examined various options and strategies. The

TABLE B-2 Energy Use for Vehicles and Fuels, Shanghai, 2000 and 2020 (kilometers per liter of fuel)

	2000	2020
Gasoline motor scooter (two-stroke)	32.1	35.5
Gasoline motor scooter (four-stroke)	44.9	49.7
Gasoline minicar	24.7	28.5
Gasoline car	10.7	10.7
Diesel car	15.8	15.8
Diesel bus	3.3	3.3
Gasoline bus	2.2	2.2

NOTES: The average generating mix for China used in calculating GHG emissions for battery electric vehicles (and rail transit) is: 78 percent coal, 15 percent hydro, 4 percent oil, 2 percent nuclear, and 1 percent natural gas. For 2020 the energy consumption, measured as joules or BTUs (British thermal units) of battery electric cars and scooters was estimated to be 10 percent less than that of comparable gasoline vehicles on an energy cycle basis, of compressed natural gas (CNG) vehicles to be 5 percent less than that of gasoline cars in terms of propulsion energy, and of hydrogen fuel cell buses to be 50 percent less than that of diesel buses.

SOURCE: For details and documentation of fuel consumption estimates, see Zhou et al. (2001) and Delucchi (1997).

final set of parameters was specified after extensive consultation among the authors and with others.

The two scenarios generated are both premised on consensus forecasts of strong, continued economic growth. If economic growth were faster or slower, vehicle ownership and energy use would be higher or lower than indicated by the scenarios. However, because this study does not address economic policy, economic variables were not considered. Even in the most conservative scenario, assuming continued economic growth, large increases will be experienced in vehicles, energy use, and GHG emissions.

Neither scenario is meant to represent (or indicate) "business-as-usual," because even that characterization is meaningless in this period of massive investments and policy shifts. Instead, these scenarios are meant to provide upper and lower bounds on likely increases in motorization and associated transportation impacts over the next 20 years.

The key parameters for the two scenarios are presented in Tables B-2 and B-3. They include population,[14] amount of motorized and non-motor-

[14] The official long-term projection is for Shanghai's official population to gradually increase from 13 to 16 million (Xia and Lu, 1999:22), equivalent to an actual population of 16 to

TABLE B-3 Key Travel and Population Parameters for Scenarios, 2000 and 2020

	2000	2020 Low	2020 High
Passengers per vehicle			
Passenger car	2.5	3.0	2.5
Scooter	1.2	1.3	1.2
Minicar	1.5	1.8	1.5
Bicycle	1.0	1.0	1.0
Bus	27	32	27
Passenger modal split by passenger-kilometer (percent)			
Gasoline cars	14	25	52
Diesel cars	0	5	0
CNG/LPG cars	1	3	0
Gasoline minicars	0	1	0
Battery and fuel cell minicars	0	4	0
Two-stroke two-wheelers	12	0	2
Electric two-wheelers	0	6	0
Four-stroke two-wheelers	0	7	5
Diesel bus	20	15	15
Gasoline bus	19	1	6
CNG bus	0	3	2
Fuel cell bus	0	2	0
Rail transit	0	16	12
Walking	7	3	3
Bicycle	27	9	3
Total	100	100	100
Population (millions)	16	18	20
Total passenger travel (ratio)	1[a]	3.4	4.2

[a] Baseline.

NOTE: CNG= compressed natural gas; LPG = liquefied petroleum gas.

18 million. But another future is plausible. Shanghai's population has increased slowly in recent years, the result of two major policies. The first is the national family planning policy that provides strong incentives for single-child families. The second policy is the local resident registration system that restricts domestic migration. In the future, as the market economy expands and the population of Shanghai ages, it may become increasingly difficult to keep poor rural residents from moving to richer cities like Shanghai. Indeed, Shanghai already houses a "floating population" numbering in the millions. Shanghai will become an increasingly strong magnet for immigration, especially for young workers from rural areas.

ized travel by mode, fuel consumption characteristics, and average vehicle occupancies.

High Motorization Scenario

This scenario is premised on market forces playing a greater role in the economy, and government playing a lesser role. It is assumed that Shanghai follows the path of fast-growing cities in Asia that have relatively high car ownership and GHG emissions for their income levels. These cities include Bangkok and Jakarta, both known for their high levels of air pollution and traffic congestion.

It also is assumed that Shanghai and the central government determine that the automotive industry will be a pillar of economic development, as conceived in the 1990s. Consumer choice is allowed to flourish, and a greater share of wealth is created and managed by the private sector.

Following this scenario of expanding private sector initiative and lessening government control, it is postulated that investments in alternative fuels founder, immigration accelerates, and investments in large public infrastructure projects slow, especially for rail transit. The car population increases fourfold, which is the mid-range forecast of the Shanghai City Comprehensive Transportation Planning Institute. Immigration exceeds official forecasts, with the overall population expanding to 20 million (compared with 16 million in 2000 and 18 million in 2020 for the low-emissions scenario). Increased immigration puts pressure on the municipal budget.

Although the demand for transport increases greatly and the multicentralization plan is well under way, government is unable to respond as it did in the 1990s. Funding for the rail transit system is suspended after only 5 of the 10 planned lines are built. Those lines that are running are popular, but daily trips by rail are only convenient for a fraction of the population. An increasing share of funds is diverted to buses, which require less capital investment than rail. Bicycle use remains high among the poor. Others walk or use buses. More bike lanes are built to serve the high demand and reduce conflicts with vehicles and buses on mixed-use roads.

The shift toward personal motor vehicles (motorcycles, scooters, and cars) accelerates for several reasons. With increased income, reduced car prices, and newly available consumer credit, many more people can purchase vehicles. Frustration over poor-quality buses and longer commutes to work lead to increased car buying. Work trips lengthen because jobs and housing become more dispersed as a result of multicentralization.

The private automobile is a symbol of wealth in Shanghai, and wealthier residents use their cars regularly despite deteriorating traffic

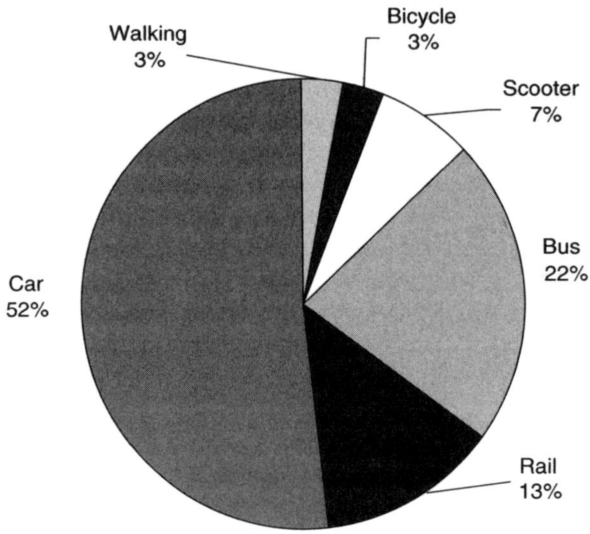

FIGURE B-3 Mode of travel in passenger-kilometers, high motorization scenario, 2020.

conditions. Dirty, inefficient two-stroke two-wheelers remain banned and are replaced by clean four-stroke and electric scooters. Not only are four-stroke engines much cleaner burning, but they consume about one-third less energy than two-stroke engines (see Table B-2). Nonetheless, their large number and intensive use results in a substantial overall increase in emissions.

The central government pursues its plan to create a strong, technologically sophisticated, domestic automotive industry with large investments from international automakers. The principal target markets are large Chinese cities such as Shanghai. Shanghai is successful in attracting a disproportionate share of the automaker investments. City officials relax vehicle taxes and other restrictions in response to the growing political clout of the local automotive industry and local motorists.

Cars increase their share of total passenger travel from 15 percent in 2000 to 52 percent in 2020. Scooters and motorcycles drop from 12 to 7 percent, bicycles from 27 to 3 percent, walking from 7 to 3 percent, and mass transit from 39 to 34 percent (see Figure B-3). These reductions in nonmotorized travel are large, but comparable with other major cities. They result from an influx of poor immigrants who cannot afford bicycles, lower population densities, and greater motorization. The net result is a 4.2-fold increase in passenger travel and over a sevenfold increase in energy use.

Low Motorization Scenario

In the low motorization scenario, Shanghai follows the path of cities such as Singapore, Tokyo, and Hong Kong. As in Singapore, government plays an active role in restraining vehicle purchases and use. But the challenge is much greater for Shanghai because it is much larger.

Motorized travel and energy use are much lower in this scenario. Rail transit plays a large role, providing high-quality service at high capacity; it serves as an attractive alternative to private vehicle use. The availability of high-quality rail transit slows the shift to personal vehicles.

Motor vehicle growth management policies, such as limitations on vehicle registrations, continue to be effective. With the population remaining relatively stable and income growing quickly, public resources are available for transportation improvements. The rail transit system is completed on schedule in 2010, and ridership is high. High-density housing is available near rail lines in satellite cities. Extensive bicycle park-and-ride lots at rail stations encourage daily commuters to use bicycles and public transport. The city invests in improved bicycle lanes and high-tech vertical bicycle parking structures at high-volume stations in dense areas.

After accession to the WTO, the central government abandons the idea of creating an automotive industry founded on conventional cars and technology. Local companies find it difficult to compete directly with automobiles from the international market. Instead, Shanghai encourages local manufacturers to build minicars, also known as city cars, and agricultural vehicles for rural areas. Disincentives are imposed on the use of larger vehicles. To provide a release for pent-up vehicle demand, Shanghai also provides seed funding, technical assistance, and parking and purchase incentives for the car-sharing organizations proliferating around the city.

Minicars become very popular. An increasing number are powered by electricity, although some have small internal combustion engines that burn gasoline and diesel fuel. Some use hybridized combinations of batteries and combustion engines. After 2010 fuel cells are used. Minicars are narrower and approximately half the length of full-size vehicles and therefore cause substantially less traffic congestion and consume much less space for parking. The lower volume of bicycle and vehicle traffic on the roads allows the remaining traffic to move faster, including buses. This gives the Shanghai bus system a strong reputation for reliability and increases ridership. The city continues with its multicentralization strategy, and satellite cities are served primarily by express bus and rail transit. Those who own city cars are able to drive them from the satellite cities to central Shanghai on roads built exclusively for small cars, motorcycles, and scooters. Because they handle only small, light vehicles, such roads take less space and cost much less to build than conventional roads. Sepa-

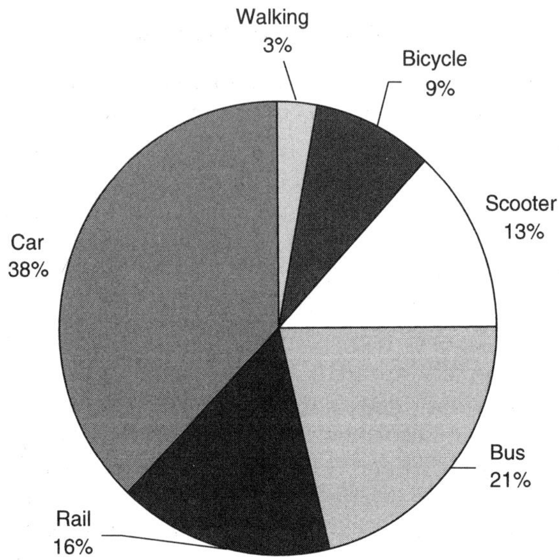

FIGURE B-4 Mode of travel in passenger-kilometers, low motorization scenario, 2020.

rate but contiguous lanes are built to higher standards at higher cost for express buses and freight trucks.

In this restrained motorization scenario, cars increase their share of travel from 15 to 38 percent between 2000 and 2020; mass transit, motorcycles, and scooters maintain their current share; and walking and bicycling drop considerably (though the absolute amount of travel by walking and bicycling stays roughly constant) (see Figure B-4). Pollution, energy use, and greenhouse gases are restrained by the use of cleaner fuels and more energy-efficient technologies. The net result in this case is a 3.4-fold increase in passenger travel and a fourfold increase in energy use.

CONCLUSION

To date, Shanghai has been highly effective in managing the demand and supply for transport during a period of explosive economic growth. Transport plans and investments have been well coordinated with larger, broader urban development plans. But strong economic growth is going to create even more pressure for personal transport and even more difficult challenges for Shanghai's leaders.

Will Shanghai side with those who believe that primary emphasis must be placed on industrial development and serving people's desires for personal transport? This policy approach supports major investments in the automotive industry, with recognition that the presence of a strong automotive industry inevitably undermines efforts to restrain vehicle use. This approach follows that of other cities and nations, including South Korea and Brazil.

Or will Shanghai maintain tight control of vehicle growth and use? Will it follow the path of cities such as Hong Kong, Singapore, and to some extent Tokyo, which restrained vehicle use?

Shanghai will experience a large increase in vehicles and energy use into the foreseeable future. The two scenarios presented here represent a plausible upper and lower trajectory of motorization and transport investments. Although both represent rapid growth, the gap between the two expands rapidly, with implications for Shanghai's economic, environmental, and social well-being.

Many factors influence Shanghai's development path, not all of which are under Shanghai's control. These factors include (1) national and local support of the domestic automotive industry; (2) the effect of China's accession to the World Trade Organization on consumer credit and vehicle availability and prices; (3) success of the multicentralization plan; (4) investments in bus and rail transit; (5) investments in road infrastructure; (6) control and pricing of parking and road use; (7) policies for motorcycles and scooters; and (8) population growth rates. The vast difference between the two scenarios suggests that there are many opportunities to influence motorization and its adverse effects.

Shanghai is already actively pursuing a broad range of transportation policies and investments. Major investments are being made in rail transit and busways; conventional roads, bridges, and tunnels, many of which support the multicentralization strategies of Shanghai; "intelligent" transportation technologies for traffic control; and new freight transport terminals and distribution centers on the outskirts of the city (important in helping divert large intercity trucks away from city streets). Existing efforts appear well organized, and policies and programs seem well integrated across various levels of city planning. But escalating demand for increased travel and vehicle ownership will create pressure for change. If the high economic and environmental cost of motorization is to be restrained, redoubled commitment to these policies is necessary now. Especially important are commitments to public transportation and restraints on car use. At the heart of these expanded initiatives is provision of a high-quality array of transportation options to travelers, including enhanced mass transit services for those otherwise inclined to shift to personal vehicles. This strategy would lead to better use of existing infrastructure and less need for infrastructure investment.

Although Shanghai is unique in many respects, virtually all of the fundamental strategies and policies examined in this case study, and even most of the specific actions, are applicable to other cities. Some of the strategies and actions proposed by the Shanghai's planning agencies are described in Box B-1. To the extent that Shanghai can restrain motorization and its adverse effects by adopting such recommendations, as in the low motorization scenario, it could serve as a model for other cities in the developing world. However, if vehicle use, energy consumption, and greenhouse gases skyrocket in Shanghai, as in the high motorization scenario, it is a signal that restraint of motorization will be virtually impossible throughout the developing world.

ACKNOWLEDGMENTS

An earlier version of this chapter was prepared for and published by the Pew Center on Global Climate Change in Washington, D.C. The Center was established by the Pew Charitable Trusts to bring a new cooperative approach and critical scientific, economic, and technological expertise to the global climate change debate. The Center is independent, nonprofit, and nonpartisan. Important contributions to this chapter from Deborah Salon and Mark Delucchi of the University of California, Davis, are gratefully acknowledged.

REFERENCES

Bose, R., D. Sperling, K. S. Nesamani, G. Tiwari, M. Delucchi, L. Redmond, and L. Schipper. 2001.Transportation in Developing Countries: Greenhouse Gas Scenarios for Delhi, India. Washington, D.C.: Pew Center on Global Climate Change.

Carsharing 2000: Sustainable transport's missing link. 2000. Journal of World Transport Policy and Practice (special issue) 5 (September).

Chang, D. T. 1999/2000. A new era for public transport development in China. Woodrow Wilson Center's China Environment Series, Issue 3. Washington, D.C.

Chen C. 1997. Shanghai urban motor vehicle emission pollution reduction strategies. Report for World Bank Project D5, July.

Delucchi, M. A. 1997. A Revised Model of Emissions of Greenhouse Gases from the Use of Transportation Fuels and Electricity. Institute of Transportation Studies, UCD-ITS-RR-97-22. University of California, Davis, November.

Energy Foundation. 2000. China eliminates 238 vehicle fees to encourage automobile ownership. China Clippings 7(June):16.

Focas, C., ed. 1998. The Four World Cities Transportation Study. London: London Research Center.

Gakenheimer, R., and D. DeLisi. 2000. Urban China Adjusting to Motorization: A Place for Car Sharing? Cooperative Mobility Program, Massachusetts Institute of Technology, May.

Gwilliam, Kenneth. 1995. Theme paper 10: Shaping the future: Getting the prices right. In China's Urban Transport Development Strategy: Proceedings of a Symposium in Beijing, November 8–10, 1995. World Bank Discussion Paper No. 352, S. Stares and Zhi L., eds. World Bank, Washington, D.C.

Institute of Tech-economics of the State Planning Commission of China (ITE-SPC). 1994. Household automobile development strategies for China. Economy Daily. October 24 (in Chinese).

Kenworthy, J. R., and F. B. Laube. 1999. An International Sourcebook of Automobile Dependence in Cities 1960–1990. Boulder: University Press of Colorado.

Logan, J. 2000. Energy and emissions for the 21st century environment. China Clippings (Energy Foundation, San Francisco, California) 8(December):30–31.

Lu X., Chen X., and Xu X. 1996. Urban Passenger Transport Planning and Urban Development. Shanghai City Comprehensive Transportation Planning Institute. Shanghai: East China Polytechnic Publishing House.

Mei A., Wu J., and Zhi L. 1998. Shanghai's land use. In The Dragon's Head: Shanghai, China's Emerging Megacity, H. D. Foster, D. C. Lai, and N. Zhou, eds. Victoria, B.C.: Western Geographical Press.

Ministry of Construction. 1994. 1993 Urban Construction Statistical Report. Beijing.

Ni C-C. 2001. Strict rules abound in Shanghai's driving test, licensing. LA Times. January 21.

Rao, D. 1999. Analysis on private car purchasing in Beijing and Shanghai. China Autos (October).

Shaheen, S., D. Sperling, and C. Wagner. 1998. Car sharing in Europe and North America: Past and future. Transportation Quarterly 52(3):35–52.

Shanghai City Comprehensive Transportation Planning Institute. 1997a. Metropolitan Transport Comparative Study—Shanghai Report for UNCRD (United Nations Centre for Regional Development).

———. 1997b. Shanghai City Transportation Study.

———. 1997c. Shanghai Transportation Report.

———. 1998. Shanghai Transportation Report.

Shanghai Highway Administration. 1999. Shanghai Metropolitan Highway Network Planning: 1999–2010. June.

Shanghai Metropolitan Highway Network Planning Agency. 1998. Shanghai Highway Network Planning, 2020.

Stares, S., and Zhi L. 1995a. Theme paper 1: Motorization in Chinese cities: Issues and actions. China's Urban Transport Development Strategy: Proceedings of a Symposium in Beijing, November 8–10, 1995. World Bank Discussion Paper No. 352. World Bank, Washington, D.C.

Stares, S., and Zhi L., eds. 1995b. China's Urban Transport Development Strategy: Proceedings of a Symposium in Beijing, November 8–10, 1995. World Bank Discussion Paper No. 352. World Bank, Washington, D.C.

State Statistical Bureau. 1998. Statistical Yearbook of Shanghai 1998. Beijing: China Statistical Publishing House.

———. 2001. China Statistical Yearbook 2001. Beijing: China Statistical Publishing House.

Stiglitz, J. E. 1997. The future of China. Speech at World Bank—International Monetary Fund Annual Meeting, Hong Kong SAR, China, September 22.

Wang B. 1999. City to initiate one-card transit. Shanghai Daily. December 28.

Weisbord, R. E. 1999. Solving China's urban crisis: China's transportation energy future. Journal of Urban Technology 6(1):89–100.

Wu W. 1999. City profile: Shanghai. Cities 16(3).

Wu Y. and Li X. 1999. Improve urban transport, protect urban environment—sustainable development for Chinese urban transportation. In China Urban Transportation for the 21st Century. Beijing: China Construction Publishing House.

Xia L. and Lu X. 1999. Review and vision of Shanghai urban transportation development. In China Urban Transportation for the 21st Century. Beijing: China Construction Publishing House.

Ying D. 1998. Shanghai's land use. In The Dragon's Head: Shanghai, China's Emerging Megacity, H. D. Foster, D. C. Lai, and Zhou N., eds. Victoria, B.C.: Western Geographical Press.

Zhou H., D. Sperling, M. Delucchi, and D. Salon. 2001. Transportation in Developing Countries: Greenhouse Gas Scenarios for Shanghai, China. Washington, D.C.: Pew Center on Global Climate Change.